Meta-Attributes and Artificial Networking

Special Publications 76

META-ATTRIBUTES AND ARTIFICIAL NETWORKING

A New Tool for Seismic Interpretation

Kalachand Sain
Priyadarshi Chinmoy Kumar

This Work is a co-publication of
the American Geophysical Union and John Wiley and Sons, Inc.

**ADVANCING EARTH
AND SPACE SCIENCE**

WILEY

Published under the aegis of the AGU Publications Committee

Matthew Giampoala, Vice President, Publications
Carol Frost, Chair, Publications Committee
For details about the American Geophysical Union visit us at www.agu.org.

The right of Kalachand Sain and Priyadarshi Chinmoy Kumar to be identified as the authors of this work has been asserted in accordance with law.

Registered Office
John Wiley & Sons, Inc., 111 River Street, Hoboken, NJ 07030, USA

Editorial Office
111 River Street, Hoboken, NJ 07030, USA

For details of our global editorial offices, customer services, and more information about Wiley products visit us at www.wiley.com.

Wiley also publishes its books in a variety of electronic formats and by print-on-demand. Some content that appears in standard print versions of this book may not be available in other formats.

Limit of Liability/Disclaimer of Warranty

Library of Congress Cataloging-in-Publication Data

Names: Sain, Kalachand, author. | Kumar, Priyadarshi Chinmoy, author. |
    John Wiley & Sons, publisher. | American Geophysical Union, publisher.
Title: Meta-attributes and artificial networking : a new tool for seismic
    interpretation / Kalachand Sain, Priyadarshi Chinmoy Kumar.
Description: Hoboken, NJ : Wiley-American Geophysical Union, 2022. |
    Includes bibliographical references and index.
Identifiers: LCCN 2021044593 (print) | LCCN 2021044594 (ebook) | ISBN
    9781119482000 (cloth) | ISBN 9781119481911 (adobe pdf) | ISBN
    9781119481768 (epub)
Subjects: LCSH: Seismology–Data processing. | Neural networks (Computer
    science)–Scientific applications. | Artificial
    intelligence–Geophysical applications.
Classification: LCC QE539.2.D36 S25 2022 (print) | LCC QE539.2.D36
    (ebook) | DDC 551.2201/13–dc23/eng/20211119
LC record available at https://lccn.loc.gov/2021044593
LC ebook record available at https://lccn.loc.gov/2021044594

Cover Design: Wiley
Cover Image: © local_doctor/Shutterstock

Set in 10/12pt Times New Roman by Straive, Pondicherry, India

SKY10034841_062322

CONTENTS

PREFACE

Surface measurements that identify and map subsurface geologic features in both onshore and offshore locations have multiple applications. These include hydrocarbon exploration; characterization of subsurface water, geothermal, and mineral resources; and understanding volcanic and seismo-tectonic processes.

Seismic data are one type of surface measurements that can be used to better understand the characteristics of the subsurface. Over the past few decades, the attributes derived from seismic data have revolutionized interpretation of geologic structures, stratigraphic features, and reservoir properties. However, as the amount of seismic data collected by scientists keeps growing, so does the task of data processing and interpretation.

High performance computing systems offer a solution to the data processing part, as they allow processing of copious amounts of data quickly. However, data interpretation can still be challenging, particularly in areas that are geologically complex and when the data volumes are large. Thus, there is a need to automate the process, accelerate the interpretation, and reduce the need for intervention by human analysts.

The purpose of this book is to introduce a new approach to seismic interpretation based on artificial networking. Several workflows have been designed to amalgamate multiple seismic attributes to compute new attributes, known as "meta-attributes" or "hybrid-attributes." The computation of meta-attributes delimits the 3-D geometry of subsurface geologic features by their seismic properties and characteristics, and thus accelerates the interpretation. The exciting part of this semi-automatic method is the fusion of human intelligence with machine intelligence over a small volume of data, followed by automatizing the process for the delineation of subsurface architecture from a huge volume of data, acquired on the surface.

The book is divided into three parts. Part I is dedicated to seismic attributes. Chapter 1 gives a brief overview of the history of seismic attributes and highlights the importance of attribute technology in exploration geophysics. Chapter 2 describes mathematical formulations of complex trace, structural, and stratigraphic attributes, and demonstrates how these can be extracted from seismic data for different purposes. Chapter 3 presents an interpretation task with ten examples of seismic cross-sections and numerics that should be of use to those new to seismic interpretation in practices.

Part II focuses on meta-attributes. Chapter 4 defines different meta-attributes that can be used for delineating various subsurface geologic features. Chapter 5 introduces the fundamentals of artificial neural networks (ANN) including historical development of comparing the mathematical neuron with the biological neuron, multi-layer perceptron (MLP) with feedforward and backpropagation algorithms for training, and types of neural networks. Chapter 6 introduces workflows for the computation of meta-attributes from seismic data.

Part III presents a series of case studies with field examples to demonstrate the application of seismic meta-attributes for automatic interpretation of 3-D geometry of subsurface geologic structures. Chapter 7 shows how gas chimneys can be picked up by the Chimney Cube Meta-attribute from reflection seismic data. Chapters 8 looks at the distribution of thin faults through the computation of the Thin Fault Cube Meta-attribute, while Chapter 9 demonstrates how a blend of Thinned Fault Cube and Fluid Cube Meta-attributes can explain the hydrocarbon fluid leakage from a network of hard-linked fault system. Chapters 10 and 11 explain how the Sill Cube Meta-attribute can capture the sill network in basins of two different geologic set ups. Chapter 12 shows an amalgamation of Sill Cube and Fluid-Cube meta-attributes that can arrest the sill complexes and fluxed-out magmatic fluids and explain their impact on forming structural folds in the overlying younger formations. Chapter 13 demonstrates the application of the Intrusion Cube Meta-attribute for interpreting a buried volcano and other intrusive elements such as the sill webs, dyke swarms, and magmatic ascent into a complex tectonic regime. Finally, Chapter 14 highlights the use of the Mass Transport Deposit Meta-attribute in deciphering the structural architecture and distribution of mass transport system from reflection seismic data.

Every chapter has a reference section to point readers to further literature, and there is a glossary at the beginning of the book explaining the main scientific terms in this field. There are also three appendices. Appendix A sheds light on the mathematical formulation of common series and transforms. Appendix B illustrates techniques of Dip-Steering. Appendix C presents answers to the interpretation and numerical tasks presented in Chapter 3. Users can make use of data sets available at www.wihg.res.in for hands-on practice of interpretation and computation of meta-attributes.

Machine learning is being applied across many fields including medical science, engineering, and astronomy. This book demonstrates its potential as a tool in the geosciences for automatic interpretation of 2-D and 3-D seismic data, a field of direct relevance to society and sustainable development. We hope that it will be of interest to specialists such as seismic interpreters in the petroleum industry as well as students and researchers who are new to seismic data analysis and interpretation.

This book is an outcome of dedicated research made by ourselves and our students on exploration seismology and their applications for subsurface imaging. We would like to convey our gratitude to our teachers and professors who nourished us from the beginning. We express deep gratitude to Prof. Harsh K. Gupta for his valuable guidance at several stages of our professional careers and personal development. We are also grateful to Prof. Sailesh Nayak for his suggestions and advice. The encouraging words of Prof. Ashutosh Sharma on the application of artificial intelligence to geosciences have inspired and motivated us in writing this book.

We thank the Government of India's Ministry of Earth Sciences, Ministry of Petroleum and Natural Gas, and the Department of Science and Technology for providing financial support to create geophysical research facilities.

Special thanks are due to Dr. Rituparna Bose at Wiley for her encouragement and valuable input from the book's inception to the final phases of writing. Thanks are due to Dr. Jun Matsushima, three other anonymous reviewers, and a member of the AGU Books Editorial Board for their constructive comments and suggestions to improve the content of our book. We also thank Dr. Jenny Lunn of AGU and Ms. Layla Harden of Wiley for their support and help at different stages of book preparation.

We acknowledge the New Zealand Petroleum Minerals, the New Zealand Ministry of Economic Development, and the dGB Earth Sciences in The Netherlands for providing the valuable data and academic license of the software.

Sincere thanks are also due to our international collaborators including Prof. Tiago M Alves (Cardiff University, UK), Dr. Kamaldeen Olakunle Omosanya (Oasisgeokonsult, Norway), Prof. Nicolas Waldmann (University of Haifa, Israel), and Prof. Qiliang Sun (China University of Geosciences), whose scientific advice have helped our students in carrying out research in applied seismology.

Finally, we acknowledge our families. From the bottom of his heart, KS affectionately thanks his wife Tumpa and son Ritwik for their continuous inspiration and love in the journey of writing this book. PCK expresses thanks to Baba and Maa. We also thank the Almighty for heavenly blessings and guidance.

Kalachand Sain
Priyadarshi Chinmoy Kumar
Wadia Institute of Himalayan Geology, India

ABOUT THE AUTHORS

Kalachand Sain is the Director of the Wadia Institute of Himalayan Geology in Dehradun, India. Previously he was Chief Scientist at the CSIR-National Geophysical Research Institute in Hyderabad, India. He has an MSc (Tech) in Applied Geophysics from IIT-Indian School of Mines, Dhanbad, and a PhD in Controlled Source Seismology from CSIR-National Geophysical Research Institute, Hyderabad. He spent time as a post-doctoral fellow at Cambridge University (UK) and Rice University (USA), and was a visiting scientist at the United States Geological Survey. His research interests include exploration of gas hydrates, imaging sub-volcanic sediments, understanding evolution of sedimentary basins and earthquake processes, and providing geotectonic implications, including the Himalayan orogeny, and glaciological and landslides hazards. He has also built expertise in travel time tomography, AVO modelling, full-waveform tomography, impedance inversion, pre-stack depth migration, seismic attenuation and meta-attributes, artificial intelligence, rock physics modelling, and interpretation of 2-D/3-D seismic data. He is a Fellow of all three Indian science academies and is the recipient of numerous medals and awards including the National Mineral Award, National Award of Excellence in Geo-sciences, J.C. Bose National Fellowship, Decennial Award & Anni Talwani Memorial Prize of Indian Geophysical Union, and Distinguished IIT-ISM Alumnus Award.

Dr. Priyadarshi Chinmoy Kumar is a Scientist at Wadia Institute of Himalayan Geology (WIHG) in Dehradun, India. He received a MSc (Tech) in Geophysics with First Class Distinction from Andhra University and Ph.D. in Science (Geophysics) from the Academy of Scientific and Innovative Research, Hyderabad. His research interests include processing and interpretation of seismic data, design and development of workflows for computation of meta-attributes, and basin studies. He has been recognized with a young scientist award by the Indian Geophysical Union, National Academy of Science, India and as an Associate of the Indian Academy of Sciences.

ACRONYMS

AAA	Anomalous Amplitude Attenuation
AFV	Average Frequency Variance
ANN	Artificial Neural Network
AOM	Ascent of Magma
AVO	Amplitude Variation with Offset
BN	Biological Neuron
BOPD	Barrels of Oil Per Day
BSR	Bottom Simulating Reflector
BS	Background Steering
BSS	Basal Shear Surface
CB	Canterbury Basin
CC	Chimney Cube
CM	Chimney Migration
CNN	Convolutional Neural Network
CTA	Complex Trace Analysis
CWT	Continuous Wavelet Transform
DHI	Direct Hydrocarbon Indicator
DS	Detailed Steering
DSDF	Dip-Steered Diffusion Filter
DSMF	Dip-Steered Median Filter
DVA	Dense Velocity Analysis
EW	East–West
E&P	Exploration & Production
FC	Fault Cube
FlC	Fluid Cube
FEF	Fault Enhanced Filter
FF	Forced Fold
FZ	Fault Zone
GC	Gas Chimney
GRNN	Generalized Regression Neural Network
HC	High Chimney
HFBZ	High Frequency Blackout Zone
HN	Hopfield Networks
HNS	Human Neural System
HP	Hydrocarbon Probability

IC	Intrusion Cube
LFBZ	Low-Frequency Blackout Zone
LVQ	Learning Vector Quantizer
LW	Long Window
MBKB	Meter Bellow Kelly Bush
MLP	Multi-Layer Perceptron
MN	Mathematical Neuron
MNN	Modular Neural Network
MP	Most Positive
MSE	Mean Square Error
MTC	Mass Transport Complex
MTD	Mass Transport Deposit
MTDC	Mass Transport Deposit Cube
MVC	Main Vent Complex
MW	Mid Window
NN	Neural Network
nRMS	Normalized RMS
NS	North–South
NZ	New Zealand
NZP&M	New Zealand Petroleum & Minerals
PD	Potential Difference
PDF	Probability Density Function
PE	Processing Element
PNN	Probabilistic Neural Network
PPS	Plio-Pleistocene Sequence
QC	Quality Check
Q-factor	Quality Factor
RBF	Radial Basis Function
REF	Ridge Enhancement Filter
RGB	Red Green Blue
RHS	Right-Hand Side
RMS	Root Mean Square
SC	Sill Cube
SD	Spectral Decomposition
SEG	Society of Exploration Geophysicists
SW	Short Window
SEG	Society of Exploration Geophysicists
S/N	Signal to Noise
SOF	Structure Oriented Filter
SOM	Self-Organizing Maps
STFT	Short-Time Fourier Transform
SV	Seismic Volume

SW	Short Window
TB	Taranaki Basin
TD	Total Depth
TFC	Thinned Fault Cube
TFL	Thinned Fault Likelihood
TWT	Two-Way-Time
UVQ	Uniform Vector Quantizer
VC	Volcanic Core
VSP	Vertical Seismic Profiling

LIST OF SYMBOLS AND OPERATORS

$A_i(t)$	Instantaneous amplitude
A_{rms}	RMS amplitude
$b_i(t)$	Instantaneous bandwidth
e_{dip}	Exaggerated dip
e_{slope}	Exaggerated slope
$f_a(t)$	Average frequency
$f_i(t)$	Instantaneous frequency
$f_{rms}(t)$	RMS frequency
f_l	Fault likelihood
γ	Exaggeration factor
H	Hilbert Transform
H^{-1}	Inverse Hilbert Transform
$\equiv$	Identical to
k	Curvature
k_{mean}	Mean curvature
k_{max}	Maximum curvature
k_{pos}	Most positive curvature
k_{neg}	Most negative curvature
k_{gauss}	Gaussian curvature
μ_{amp}	Average of the amplitudes
φ	Azimuth
p	inline dip component
$p(\varphi)$	Apparent dip of the inline component measured at an azimuthal angle
q	xline dip component
$q(\varphi)$	Apparent dip of the xline component measured at an azimuthal angle
$q_a(t)$	Average quality factor
$q_i(t)$	Quality factor or instantaneous quality factor
R	Radius of curvature
r_{pseudo}	Pseudo relief
ρ_x	Inline lag cross-correlation
ρ_y	Xline lag cross-correlation
$\sigma_a(t)$	Average decay rate
σ_{amp}	Standard deviation of amplitudes

$\sigma_i(t)$	Instantaneous decay rate
$\sigma_r(t)$	Relative amplitude
$\langle \bullet \rangle_s$	Structure-oriented averaging
$S(t)$	Sweetness
S_m	Similarity
S_{dip}	Dip-steered similarity
ψ_s	Strike of a surface
θ_d	Dip angle or true dip of a surface
$\theta_i(t)$	Instantaneous phase
θ_x	Apparent dip in x-direction
θ_y	Apparent dip in y-direction
u	Amplitude in a seismic cube
$x(t)$	Real component of seismic trace
$y(t)$	Quadrature component seismic trace

GLOSSARY

Activation function	A mathematical function that maps the output of the neural network in terms of binary values.
Amplitude	The magnitude values of the seismic trace or trace envelope.
Amplitude change	The change in seismic amplitude over an interval in a given direction.
Azimuth	The angle measured clockwise from the geographic north in the direction of the maximum downward dip or slope.
Backpropagation	A process of computing the error in prediction to the actual in the backward direction.
Bandwidth	The breadth of the frequency power spectrum of a waveform.
Bright spot	A local high-amplitude seismic anomaly that shows the presence of hydrocarbons. Also known as the direct hydrocarbon indicator.
Curvature	The degree of curvedness of a surface, and is defined as the curvature, i.e. how or to what extent a surface bends or curves.
Chimney (or gas chimney)	Vertical disturbances observed on seismic data. Such features are associated with chaotic reflections where amplitudes are weaker.
Co-rendering	A process in which attribute maps are superimposed and color scales are adjusted to improve visualization of subsurface geologic features.
Coherency	A measure of lateral change in seismic response caused by the variation in structure, stratigraphy, or lithology.
Data conditioning	A process of improving signal-to-noise (s/n) such that the seismic data can be readily interpreted.
Depth slice	A process of slicing a 3D seismic cube at a particular depth level.
Detailed steering cube	A steering cube that is obtained by applying mild filtering.
Dip-steering	A process of estimating dip-azimuth information at each sample location of a seismic trace. The process generates a steering cube, called the dip-steering volume or dip-azimuth volume.

Example locations	Places that consist of targets and non-targets.
Faults	Discontinuous structures that are generally associated with reflector terminations, vertical disturbances, and breaks in seismic reflectors.
Fault dip	The angle that the fault plane makes with the horizontal.
Fault throw	The vertical separation of a layer generated due to faulting.
Feedforward	The process of feeding a neural network in the forward direction.
Filtering	A process of removing unwanted information from the data.
Frequency shadow	Loss or wash-out of higher frequencies leaving behind a lower frequency that generates a shadow beneath the reservoir.
Gas chimney	Upward migration and escaping of accumulated gas and appears as a wash-out zone in seismic sections.
Instantaneous amplitude (IA)	Change in amplitude at a given instant of time, or it represents the magnitude of the sinusoid at a given time that represents a seismic trace.
Instantaneous phase (IP)	Change in phase at a given instant of time.
Instantaneous frequency (IF)	Change in frequency at a given instant of time.
Magmatic Sill	Tabular intrusive rocks with concordant surfaces showing concave upwards cross-sectional geometries and discordant limbs.
Mass Transport Deposit	Unconsolidated sediments transported into the deep water environment under the influence of gravitational force due to slope failure or slope instability.
Meta-attribute	A hybrid attribute generated by combining a set of other seismic attributes using a neural-based approach.
Migration	A process carried out to move dipping reflectors to their correct position to produce accurate images of the subsurface.
Misclassification	A quality control parameter used to understand the wrong prediction made during classification.
Multi-layer	A network consisting of several layers.
Multi-layer perceptron	A network arrangement in which nodes or neurons in each layer are fully connected to the nodes or neurons of the consecutive layers.
Overtraining	A process in which neural training becomes non-universal, meaning the network fails to differentiate between the target and non-targets.

Perceptron	The building elements in a layer of a neural network.
Phase	The relative position along a seismic waveform that is independent of amplitude.
RGB blending	A process of blending red, green, and blue colors to improve visualization of geologic features from data.
Reflection	Seismic wave that gets reflected from an interface where there is a change in acoustic impedance.
Relative amplitude change	The change in the instantaneous amplitude.
Seismic discontinuity	A break in the continuity of a seismic reflector.
Seismic horizontal resolution	The ability of distinguishing two closely spaced reflectors laterally in a seismic section.
Seismic vertical resolution	The ability of distinguishing two closely spaced reflectors vertically in a seismic section.
Slope	The ratio between the change in depth of reflector over a change in horizontal distance.
Soma	The central part of the human neural system that contains solution rich in potassium ions rather than sodium ions.
Structural attributes	A set of attributes that quantify structural properties of geologic features, e.g., slope, azimuth, dip, curvature, relief, and discontinuity.
Stratigraphic attributes	A set of attributes that elaborates the reflection patterns related to the stratigraphy.
Sigmoid function	An s-shaped mathematical function that is continuous, differentiable, and monotonically increasing.
Similarity	Similar nature in two segments of seismic traces.
Supervised Learning	A learning approach in which the perceptron or network is made to learn from available examples.
Steering cube	A steering cube that is obtained by applying coarser filtering.
Sweetness	An attribute, defined as the ratio of RMS amplitude to square root of instantaneous frequency, associated with oil and gas deposits.
Synapses	A structure in the human nervous system that permits a neuron to pass an electrical or chemical signal to other neuron.
Training	A process through which a neural network is taught.

Time slice	A process of slicing a 3D seismic cube at a particular time level.
Wavelength	The distance between two successive peaks or troughs in a seismic wave.
Weights	Measures the strength of the connection of neurons in a layer with those in other layers.
Unconformity	A break in the geological record.
Unsupervised learning	A learning approach that teaches itself regarding structures or pattern present within the data.

ADDITIONAL RESOURCES

Students, teachers, or instructors can download sample data sets from our institute server for hands-on practice. Note that the files are in SGY format. Software such as SeisSee, Seismic Trace Viewer (STV), Open Inventor SEG Y reader, GSEGYView, kogeo or SEGY Scout is needed to open these files and visualize the data. Separate seismic interpretation softwares such as PetrelTM, LandmarkTM, JasonTM, HRSTM, SeisEarthTM, DUG InsightTM, PaleoS-canTM or OpendTectTM is then needed for interpreting the SEGY data files.

To access the files, go to http://14.139.225.217/ai_book_data/ai_book_data.zip. When prompted enter user1 as username and wihg248))! as password. The data will be downloaded as a zipped file. If you experience any difficulties accessing the files, please contact us at kumarchinmoy@gmail.com or chinmoy@wihg.res.in.

Part I
Seismic Attributes

1

AN OVERVIEW OF SEISMIC ATTRIBUTES

Seismic attributes play a vital role in the interpretation of subsurface geological features such as faults, fractures, folds, channels, diapirs and reefs, and in inferring dynamic and static properties of subsurface reservoirs. Hence, understanding of attributes and their extraction from surface seismic data are crucial for the illumination of subsurface structures and properties. This chapter provides an overview of seismic attributes, their historical evolution, and fundamental characteristics or properties that can differentiate objects and their subsurface disposition.

1.1. Introduction

Seismic attributes have been used over the past few decades to infer subsurface properties and geologic features from reflection seismic data. These attributes illuminated geologic features, revealed structural architecture, and quantified specific physical properties. The analysis of seismic attributes plays a pivotal role in petroleum exploration through the delineation and interpretation of subsurface faults, fractures, folds, channels, diapirs, reefs (Figure 1.1), etc. These subsurface features act as traps for hydrocarbon accumulation and help in predicting dynamic and static characteristics of subsurface reservoirs (Chopra & Marfurt, 2007). Since inception, a large number of attributes have been generated from seismic data, which have been efficiently utilized to describe and delimit different geologic targets of interests.

Meta-Attributes and Artificial Networking: A New Tool for Seismic Interpretation,
Special Publications 76, First Edition. Kalachand Sain and Priyadarshi Chinmoy Kumar.
© 2022 American Geophysical Union. Published 2022 by John Wiley & Sons, Inc.
DOI: 10.1002/9781119481874.ch01

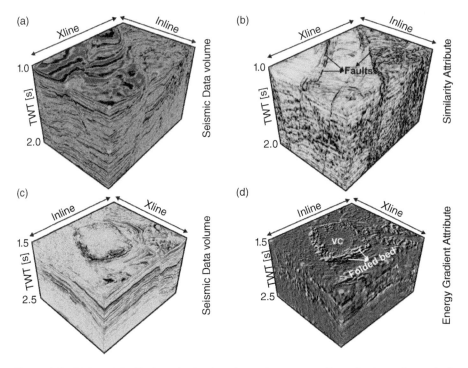

Figure 1.1   Volumetric display of seismic cube and corresponding seismic attributes (a–b and c–d), demonstrating their efficiency in describing different subsurface geological features (after Kumar and Sain, 2018; Kumar et al., 2019; VC: Volcanic Core).

Seismic attributes are used to extract information from the pre-stack or post-stack data i.e., gathers or volumes. Pre-stack attributes treat seismic data as records of seismic reflections that are associated with the P- and S-impedances, P- and S- wave velocities, amplitude variation with offset (AVO), attenuation, anisotropy, AVO intercept, and gradients. However, post-stack attributes consider seismic data to be a representation of Earth's subsurface image that includes a large family of attributes, e.g., complex trace attributes, interval attributes, horizon-based attributes, time-frequency attributes, and waveforms (Barnes, 2016; Al-Shuhail et al., 2017). The attributes, as a whole, follow a unified characteristic, based on which maximum information about a target can be extracted and subsurface architecture of a geologic body can be inferred from data. Thus, they act as filters, designed in such a way as to illuminate the properties of interest by setting aside those which are of no interest. Though both domains divide the family of seismic attributes by assigning different means of usage, attribute analysis has received the utmost attention in image processing and enhancement, which aims to extract valuable subsurface information from surface data.

This chapter shows how the seismic attributes have evolved and are used for enhancing characteristic properties stored within the seismic data. Most of these properties form the basis for designing and formulating seismic attributes for subsurface interpretation from surface data.

1.2. Historical Evolution of Seismic Attributes

The 1920s marked the beginning of field reflection seismic experiments, which mapped subsurface geological structures by identifying reflections and converting the arrival times into corresponding depths. This practice continued through the 1950s until the late 1960s. The advent of Analog-to-Digital (A/D) conversion techniques facilitated seismic data processing on digital computers. However, this digital revolution still did not stream interpretation practices by geophysicists who could correctly map subsurface geology from data. Rummerfeld (1954) was one of those visionaries who could qualitatively use reflection characteristics to interpret subsurface stratigraphy from seismic data. This event triggered the enthusiasm of geophysicists for critically understanding the properties from seismic data for inferring subsurface geological structures. Koefoed (1955) laid down his interpretational insights into signal processing from amplitude variations with offsets (AVO), which led to the interpretation of subsurface lithological properties. Thereafter, several geophysicists (Merlini, 1960; Savit, 1960) documented their pioneering works for the interpretation of seismic reflection data. The digital revolution brought about significant changes in signal processing and quality (Yilmaz, 2001). Slowly, the interpretation of seismic reflections in inferring subsurface stratigraphy became a routine job.

The late 1960s witnessed a significant discovery from such practices. Recognition of "bright spots" from seismic data by Soviet geophysicists opened up a new era in interpretation strategies. These anomalous features aroused interest in the direct detection of hydrocarbons, geared up seismic explorationists, fascinated many seismic contractors, and led to successful exploration cases in the Gulf of Mexico. The bright spot detection through the 1970s captivated the interpretation community, making the first seismic attribute "reflection amplitude."

Reflection strength (Figure 1.2) is a classic example of amplitude attribute, designed by Anstey (1972).

Today this attribute is considered the most important and powerful of all existing seismic attributes. Once the amplitude phenomena led to the direct search for hydrocarbons, researchers immediately became curious about the frequency characteristic of the signal. When seismic waves propagated through a gas

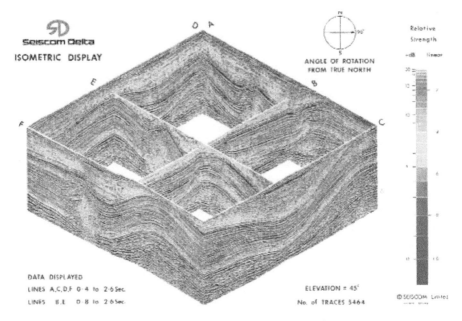

Figure 1.2   Isometric display of reflection strength attribute (after Anstey, 1972).

reservoir, it was observed that the signal suffered from higher frequency being washed out, leaving behind the lower frequency, thereby causing a shadow. Sheriff (1975) called this "Frequency Shadow Effect", which directly indicates the presence of a gas reservoir. Dobrin & Savit (1960) documented that such attenuation could be used for the quantification of rock quality factor. Anstey (1977, 2001, 2005) developed a novel procedure for attribute analysis, in which he demonstrated the use of color displays for analyzing seismic attributes. The "Complex Trace Analysis" made its first appearance in the 1976 annual meeting of the Society of Exploration Geophysicists (SEG) in the form of seminar papers by Taner, Sheriff and his group, which later on became the first masterpiece of work. Based on this concept, Taner & Sheriff (1977) and Taner et al. (1979) developed five attributes: instantaneous amplitude, instantaneous phase, instantaneous polarity, instantaneous frequency, and weighted average frequency. While these developments were gaining pace, Peter Vail and his group formulated the principles of seismic stratigraphy. Being inspired by the pioneering works of Taner and his group, the complex trace attributes found their place in explaining the seismic properties related to the stratigraphy. Several other attributes, e.g., root mean square amplitude, zero-crossing frequency, and cosine of phase, which were observed in early 1980s, were also regarded to be more comprehensible substitutes for complex trace seismic attributes. On the advent of 3D seismic data and the use of computer systems during the mid-80s, attribute maps and their analysis came to light. The first attribute maps were the simple attributes extracted from seismic amplitude volume (Denham & Nelson, 1986), horizon attributes, and horizon-guided interval attributes (Bahorich and Bridges, 1992; Dalley et al., 1989; Hoetz & Waters, 1992; Rijks & Jauffred, 1991). Bahorich & Farmer (1995) further developed 3D discontinuity attributes, which, when displayed through time and horizon slices, distinctly revealed faults, salt domes, and meandering channels. This caused curiosity among interpreters and led to the revitalization of seismic attribute analysis. This development opened an avenue for assessing other 3D properties such as dips, azimuths, curvatures, parallelism, etc. (Marfurt et al., 1999; Oliveros & Radovich, 1997; Randen et al., 2000; Taner, 2001). Mapping of thin beds and channel deposits from seismic data led to the development of spectral decomposition, which also made a breakthrough in seismic attribute analysis (Gridley & Partyka, 1997). Such an approach, coupled with tuning thickness analysis, led to a step forward for the quantitative application of seismic attributes.

Multi-attribute analysis gained importance and found a place in routine applications for seismic data interpretation. It is notable that the supervised methods were more preferred, as these methods could be trained to produce results of geological importance (Aminzadeh & de Groot, 2004; Hampson et al., 2001; Meldahl et al., 2002; Nikravesh et al., 2003). However, extraction of meaningful geology through unsupervised methods remained challenging. To date, a plethora of seismic attributes have been developed to capture the responses from subsurface geologic features for meaningful interpretation. Seismic attributes and their evolution have been very important in quenching the thirst of interpreters.

1.3. Characteristics of Seismic Attributes

Seismic attributes possess interesting characteristics that make them unique tools for the interpretation of subsurface geology.

They operate as filter: Seismic attributes act as filters in which they have the ability to highlight the desired component of data by removing the unwanted elements.

They perform qualitative and quantitative operations: Attributes can be used to divulge geological complexities, such as the structural and stratigraphic architecture of the subsurface, and to quantify reservoir properties, such as porosity, saturation, and permeability.

They convey geological or geophysical implications: For example, the discontinuity attribute signifies high dissimilarity or low continuity in a geophysical sense, which is associated with the faults or fractures in geological terms.

1.4. A Glance at Seismic Characteristics

Barnes (2016) defines seismic properties/characteristics as geological, geophysical, or mathematical. Seismic data possess several characteristics and seismic attributes, which are commonly used in describing the geological features. A few seismic attributes are amplitude, phase, frequency, bandwidth, amplitude change, tuning thickness, and waveform. Some important characteristics are dip, azimuth, curvature, discontinuity, and parallelism. Below is a short description of these properties.

1.4.1. Amplitude

The amplitude (Figure 1.3) is the most crucial seismic property and plays an important role in formulating many other attributes.

The amplitude attribute is defined as the magnitude values of a seismic trace or trace envelope. The most commonly used amplitude attributes are the reflection strength, root mean square (RMS) amplitude, average amplitude, maximum amplitude, and trace envelope. The reflection strength is generally computed through complex trace analysis, and is synonymous with trace envelope and instantaneous amplitude.

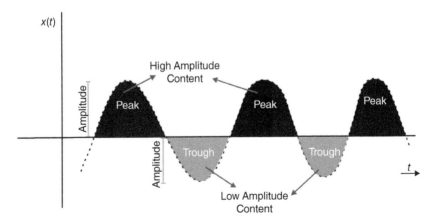

Figure 1.3   Display of seismic amplitude for a sinusoidal wave.

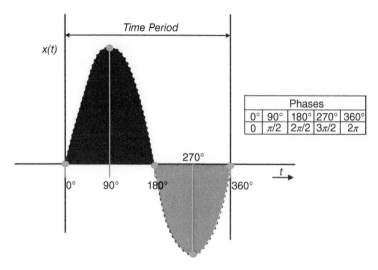

Figure 1.4 Phase measures along a seismic waveform x(t).

1.4.2. Phase

Phase (Figure 1.4) refers to the relative position along a seismic waveform and is independent of amplitude. The common seismic attributes derived from phase are the instantaneous phase, response phase, and apparent polarity (Barnes, 2016).

1.4.3. Frequency

The number of sinusoidal cycles that occur along a seismic waveform within one second of time is called frequency. The common frequency attributes include instantaneous frequency, zero frequency, average spectral frequency, RMS frequency, and tuning frequency. The frequency attribute is most commonly applied in measuring bed thickness and seismic attenuation.

1.4.4. Bandwidth

Bandwidth refers to the breadth of the frequency power spectrum of a waveform. For a given seismic trace, the bandwidth is a function of change in frequency and amplitude along the trace (Barnes, 2016). Bandwidth attributes are used to discriminate stratigraphic features.

1.4.5. Amplitude Change

Amplitude change is defined as the change in seismic amplitude over an interval in a given direction. Geologic faults or edges of channels undergo a significant change in amplitude (for example, fault zones are associated with amplitude loss). The amplitude change attribute provides detailed information from the data.

1.4.6. Slope, Dip, and Azimuth

The orientations of seismic reflectors are determined by the dip and azimuth of a reflector and serve to formulate the dip and azimuth seismic attributes. The slope is defined as the ratio of the change in depth of reflection over a change in the horizontal distance. The arctangent of the slope outputs the dip, which is defined as the angle in degrees that a seismic reflection makes with the horizontal, and is expressed in microseconds per meter. However, to obtain the values in degrees, the dip must be estimated using a conversion velocity. The slope is considered a geophysical property, whereas the dip is a geological property. Azimuth is the angle measured in the clockwise from the geographic north in the direction of maximum downward dip or slope. Figure 1.5 (a, b) demonstrates the measurement of slope and dip components in 2D and 3D.

1.4.7. Curvature

Curvature is defined as the degree of curvedness of a surface, i.e. how or to what extent a surface bends or curves (Figure 1.6).

The curvature attribute is also defined as the rate of change of dip and azimuth along with a seismic reflection or an identified seismic horizon (Barnes, 2016). Roberts (2001) defines a complicated set of curvature properties, namely the Gaussian curvature, mean curvature, maximum and minimum curvature, most

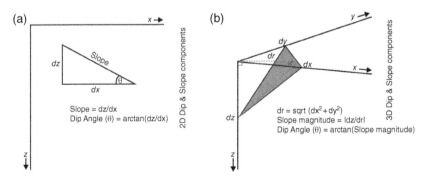

Figure 1.5   Dip and slope in (a) 2D and (b) 3D measurements.

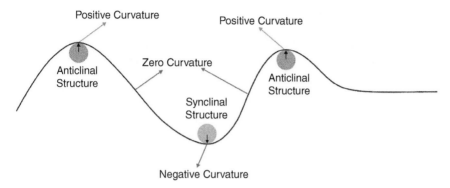

Figure 1.6   Display of curvature measures for different kinds of geologic structures.

positive and most negative curvature, dip curvature, strike curvature, curvedness, and shape index. Most positive and most negative curvature measures demonstrate a wide range of ground applications, e.g., submarine channel interpretation, identifying displaced fault blocks, fracture interpretation, etc. By sign convention, anticlines or reflection bumps are associated with positive curvature, and those of synclines or bowl-shaped features are related to negative curvature (Figure 1.6). The upthrown portion of a formation displaced by a fault has a positive curvature, and the downthrown part is associated with the negative curvature. However, Rich (2008) argued that these set of curvatures cannot measure true curvatures. The curvature measures are very important for fracture delineation (Lisle, 1994), as fractures mostly occur in anticline tops, synclinal bottoms, and at flexures. High curvature values may not indicate fractures but could ascertain fracture-prone areas or where fractures are more likely to form.

1.4.8. Seismic Discontinuity

Seismic discontinuity refers to a break in the continuity of seismic reflections. It is often referred to as "coherence" or "similarity" and is widely used for the interpretation of geological features, e.g., faults, diapirs, pinch-outs, channel belts, noise, and artefacts (Figures 1.7 and 1.8).

It is opposite to seismic continuity, where continuity refers to the degree to which seismic reflectors exhibit consistent amplitude and phase. Seismic discontinuity is a structural or geometrical attribute.

1.5. Summary

This chapter has presented a brief overview of seismic attributes from their historical evolution and outlined different characteristics that form the basis for

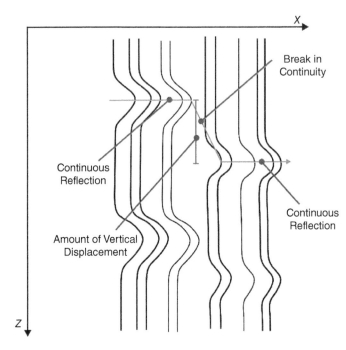

Figure 1.7   Seismic discontinuity, observed within the seismic trace, which is defined as break in seismic reflectors. This is a characteristic property of discontinuous structures used for interpretation from seismic data.

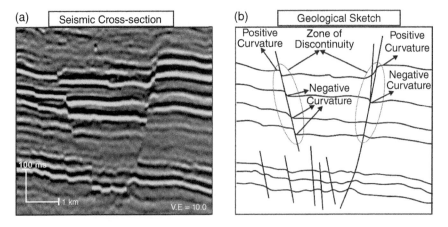

Figure 1.8   (a) Seismic discontinuity and curvature property on seismic amplitude section that has been explained through (b) a sketch for interpretation of geologic faults.

their formulation and computation. Based on these ideas, the next chapter explores different seismic attributes that have been widely used for the interpretation of subsurface geologic features from seismic data.

References

Al-Shuhail, A. A., Al-Dossary, S. A., & Mousa, W. A. (2017). *Seismic data interpretation using digital image processing*. John Wiley & Sons. https://doi.org/10.1002/9781119125594

Aminzadeh, F., & de Groot, P. (2004). Soft computing for qualitative and quantitative seismic object and reservoir property prediction. Part 1: neural network applications. First Break, *22*(3). https://doi.org/10.3997/1365-2397.22.3.25812

Anstey, N. (1972). Seiscom'72 (Seiscom Limited internal report).

Anstey, N. A. (1977). Seismic interpretation: The physical aspects: International Human Resources Development Corp., http://dx.doi.org/10.1007/978-94-015-3924-1.

Anstey, N. (2001). Snapshots from a lifetime in geophysics. In A. McBarnet (Ed.), *EAGE, 1951–2001: Reflections on the first 50 years* (pp. 3–10). Blackwell Science.

Anstey, N. (2005). Attributes in color: the early years: CSEG Recorder, *30*(3), 12–15.

Bahorich, M. S., & Bridges, S. R. (1992). Seismic sequence attribute map (SSAM). Paper presented in in *SEG Technical Program Expanded Abstracts 1992* (pp. 227–230). Society of Exploration Geophysicists. https://doi.org/10.1190/1.1822047

Bahorich, M., & Farmer, S. (1995). 3-D seismic discontinuity for faults and stratigraphic features: The coherence cube. Leading Edge, *14*, 1053–1058. https://doi.org/10.1190/1.1437077

Barnes, A. E. (Ed.). (2016). *Handbook of poststack seismic attributes*. Society of Exploration Geophysicists. https://doi.org/10.1190/1.9781560803324

Chopra, S., & Marfurt, K. J. (2007). *Seismic attributes for prospect identification and reservoir characterization*. Society of Exploration Geophysicists and European Association of Geoscientists and Engineers. https://doi.org/10.1190/1.9781560801900

Dalley, R. M., Gevers, E. C. A., Stampfli, G. M., Davies, D. J., Gastaldi, C. N., Ruijtenberg, P. A., & Vermeer, G. J. O. (1989). Dip and azimuth displays for 3D seismic interpretation. First Break, *7*(3), 86–95. https://doi.org/10.3997/1365-2397.2007031

Denham, J. I., & Nelson Jr, H. R. (1986). Map displays from an interactive interpretation. Geophysics, *51*(10), 1999–2006. https://doi.org/10.1190/1.1442055

Dobrin, M. B., & Savit, C. H. (1960). *Introduction to geophysical prospecting* (Vol. *4*). New York: McGraw-Hill. https://doi.org/10.1093/gji/3.3.378

Gridley, J., & Partyka, G. (1997). Processing and interpretational aspects of spectral decomposition. Paper presented in *SEG Technical Program Expanded Abstracts 1997* (pp. 1055–1058). Society of Exploration Geophysicists. https://doi.org/10.1190/1.1885566

Hampson, D. P., Schuelke, J. S., & Quirein, J. A. (2001). Use of multiattribute transforms to predict log properties from seismic data. Geophysics, *66*(1), 220–236. https://doi.org/10.1190/1.1444899

Hoetz, H. L. J. G., & Watters, D. G. (1992). Seismic horizon attribute mapping for the Annerveen Gasfield, the Netherlands. First Break, *10*(2). https://doi.org/10.3997/1365-2397.1992003

Koefoed, O. (1955). On the effect of Poisson's ratios of rock strata on the reflection coefficients of plane waves. Geophysical Prospecting, *3*(4), 381–387. https://doi.org/10.1111/j.1365-2478.1955.tb01383.x

Kumar, P. C., & Sain, K. (2018). Attribute amalgamation-aiding interpretation of faults from seismic data: An example from Waitara 3D prospect in Taranaki basin off New Zealand. Journal of Applied Geophysics, *159*, 52–68. https://doi.org/10.1016/j.jappgeo.2018.07.023

Kumar, P. C., Omosanya, K. O., & Sain, K. (2019). Sill Cube: An automated approach for the interpretation of magmatic sill complexes on seismic reflection data. Journal of Marine and Petroleum Geology, *100*, 60–84. https://doi.org/10.1016/j.marpetgeo.2018.10.054

Lisle, R. J. (1994). Detection of zones of abnormal strains in structures using Gaussian curvature analysis. AAPG Bulletin, *78*, 1811–1819. https://doi.org/10.1306/A25FF305-171B-11D7-8645000102C1865D

Marfurt, K. J., Sudhaker, V., Gersztenkorn, A., Crawford, K. D., & Nissen, S. E. (1999). Coherency calculations in the presence of structural dip. Geophysics, *64*(1), 104–111. https://doi.org/10.1190/1.1444508

Meldahl, P., Najjar, N., Oldenziel-Dijkstra, T., & Ligtenberg, H. (2002). Semi-automated detection of 4D anomalies. Paper presented in *64th EAGE Conference & Exhibition* (pp. cp-5). European Association of Geoscientists & Engineers. https://doi.org/10.3997/2214-4609-pdb.5.P315

Merlini, E. (1960). A new device for seismic survey equipment: Geophysical Prospecting, *8*(1), 4–11. https://doi.org/10.1111/j.1365-2478.1960.tb01483.x

Nikravesh, M., Zadeh, L. A., & Aminzadeh, F. (Eds.), (2003). *Soft computing and intelligent data analysis in oil exploration*. Elsevier, Amsterdam.

Oliveros, R. B., & Radovich, B. J. (1997). Image-processing display techniques applied to seismic instantaneous attributes over the Gorgon gas field, North West Shelf, Australia. Paper presented in SEG *Technical Program Expanded Abstracts 1997* (pp. 2064–2067). Society of Exploration Geophysicists. https://doi.org/10.1190/1.1885862

Randen, T., Monsen, E., Signer, C., Abrahamsen, A., Hansen, J.O., Sæter, T. & Schlaf, J. (2000). Three-dimensional texture attributes for seismic data analysis. 70th Annual International Meeting, SEG, Expanded Abstracts, 668–671. https://doi.org/10.1190/1.1816155

Rich, J. (2008). Expanding the applicability of curvature attributes through clarification of ambiguities in derivation and terminology. Paper presented in *SEG Technical Program Expanded Abstracts 2008* (pp. 884–888). Society of Exploration Geophysicists. https://doi.org/10.1190/1.3063782

Rijks, E. J. H., & Jauffred, J. C. E. M. (1991). Attribute extraction: An important application in any detailed 3-D interpretation study. The Leading Edge, *10*(9), 11–19. https://doi.org/10.1190/1.1436837

Roberts, A. (2001). Curvature attributes and their application to 3D interpreted horizons. First Break, *19*(2), 85–100. https://doi.org/10.1046/j.0263-5046.2001.00142.x

Rummerfield, B. F. (1954). Reflection quality, a fourth dimension. Geophysics, *19*(4), 684–694. https://doi.org/10.1190/1.1438038

Savit, C. H. (1960). Preliminary report: A stratigraphic seismogram. Geophysics, *25*(1), 312–321. https://doi.org/10.1190/1.1438697

Sheriff, R. E. (1975). Factors affecting seismic amplitudes. Geophysical Prospecting, *23*(1), 125–138. https://doi.org/10.1111/j.1365-2478.1975.tb00685.x

Taner, M. T., & Sheriff, R. E. (1977). Application of amplitude, frequency, and other attributes to stratigraphic and hydrocarbon determination: Section 2. Application of seismic reflection configuration to stratigraphic interpretation. In C.E. Payton (Eds.), *Seismic stratigraphy: Applications to hydrocarbon exploration: AAPG Memoir 26* (pp. 301–327). AAPG.

Taner, M. T., Koehler, F., & Sheriff, R. E. (1979). Complex seismic trace analysis. Geophysics, *44*(6), 1041–1063. https://doi.org/10.1190/1.9781560801580.fm

Taner, M. T. (2001). Seismic attributes. CSEG Recorder, *26*(7), 49–56.

Yilmaz, Öz (2001). Seismic data analysis: Processing, inversion, and interpretation of seismic data. S. M. Doherty (Ed.), Society of Exploration Geophysicists, Tulsa, OK, USA. https://doi.org/10.1190/1.9781560801580.fm

2

COMPLEX TRACE, STRUCTURAL, AND STRATIGRAPHIC ATTRIBUTES

This chapter describes the seismic amplitudes and phases extracted by complex trace analysis. Several other complex attributes can be derived from these fundamental attributes. The mathematical formulation of complex attributes and their application to seismic data is described. The usage of structural and stratigraphic attributes for the interpretation of subsurface features is explained. This chapter also provides the definitions and mathematical framework of complex attributes that are fundamental cornerstones of subsequent chapters. It highlights pitfalls that may be encountered during an interpretation.

2.1. Introduction

Complex trace analysis (CTA) is important, because it separates the amplitude and phase information for meaningful interpretation of subsurface from surface data. The CTA results in two separable attributes: (i) instantaneous amplitude and (ii) cosine of instantaneous phase. Other attributes that are derived from the amplitude and phase attributes are the instantaneous frequency, relative amplitude, bandwidth, dip, and azimuth. Structural and stratigraphic attributes are used to capture different geologic features such as the structural and stratigraphic characters of subsurface. For example, dip-azimuth and coherency attributes measure the orientations and breaks of seismic reflectors, which help in identifying the geologic faults from seismic data. Similarly, the subsurface

Meta-Attributes and Artificial Networking: A New Tool for Seismic Interpretation,
Special Publications 76, First Edition. Kalachand Sain and Priyadarshi Chinmoy Kumar.
© 2022 American Geophysical Union. Published 2022 by John Wiley & Sons, Inc.
DOI: 10.1002/9781119481874.ch02

depositional environment can be deciphered by analyzing reflection patterns. These are explained in more detail in subsequent sections.

2.2. Complex Trace Attributes: Mathematical Formulations and Derivations

Consider a seismic trace $x_i(t)$ that can be expressed as a product of instantaneous amplitude $A(t)$ and cosine of instantaneous phase $\theta_i(t)$ as:

$$x_i(t) = A_i(t) \cos (\theta_i(t)).$$

However, this equation does not individually define the amplitude and phase components. Applying a Hilbert Transform to this equation changes the cosine wave into a sine wave making a −90° phase change to the seismic trace. Thus, the resultant quadrature trace becomes:

$$y_i(t) = A_i(t) \sin (\theta_i(t)).$$

Now the instantaneous amplitude $A_i(t)$ and phase $\theta_i(t)$ can be explicitly derived as:

$$A_i(t) = \sqrt{x_i(t)^2 + y_i(t)^2} \tag{2.1}$$

and

$$\theta_i(t) = \arctan \frac{y_i(t)}{x_i(t)}. \tag{2.2}$$

The instantaneous amplitude (Eq. 2.1) is an amplitude, commonly referred to the "trace envelope" or "reflection strength". It is defined as the change in amplitude at a given instant of time or, in other words, it represents the magnitude of the sinusoid at a given time that best matches the seismic trace. The square of the instantaneous amplitude results in the instantaneous power, which is often used as a weighting function for average attributes (Barnes, 2016) to highlight amplitude anomalies from seismic data (Figures 2.1, 2.2, and 2.3).

The instantaneous phase (Eq. 2.2) is an angular measure in degrees and is defined as the change in phase at a given instant of time. It represents the phase of the sinusoid at a given time that best matches the seismic trace (Barnes, 2016). For peaks of a trace, the instantaneous phase is 0° and the troughs are characterized by 180°. Instantaneous phase is mathematically discontinuous and brings out a sawtooth effect in color displays. Hence, to circumvent this, interpreters use cosine of the phase, which is continuous and appears to increase with gain in seismic data.

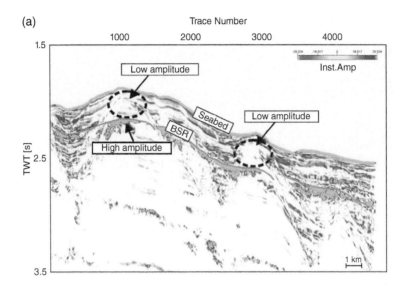

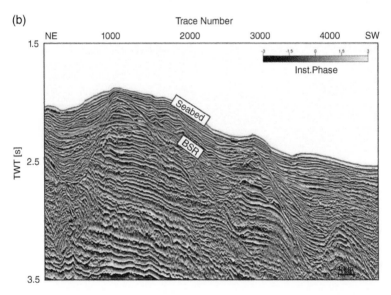

Figure 2.1 (a) Instantaneous amplitude and (b) instantaneous phase demonstrating the anomalies pertaining to gas hydrates and free gas-bearing zones in seismic data from Mahanadi Basin, eastern Indian offshore. Source: Kumar et al. (2019a), Figure 08[p.08]/ with permission of Springer Nature.

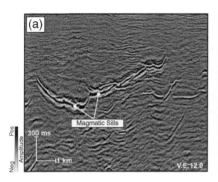 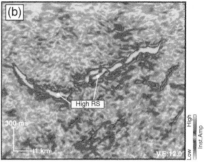

Figure 2.2 (a) Amplitude section showing the presence of magmatic sills in seismic data from Vøring Basin, offshore Norway. (b) Computed instantaneous amplitude, demonstrating high reflection strength within the magmatic sills (after Kumar et al., 2019b).

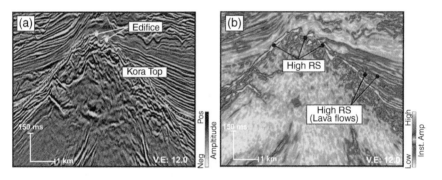

Figure 2.3 (a) Amplitude section showing the presence of a buried volcano (the well-known Kora volcano) in seismic data from Taranaki Basin, offshore NZ. (b) Computed instantaneous amplitude revealing high reflection strength along the volcanic top and lava flows (after Kumar et al., 2019c).

2.3. Other Derived Complex Trace Attributes

2.3.1. Instantaneous Frequency

The instantaneous frequency is defined as the time derivative of the instantaneous phase, and is mathematically expressed as:

$$f_i(t) = \frac{1}{2\pi} \frac{d}{dt} \theta_i(t). \tag{2.3}$$

For a given instant of time, the instantaneous frequency represents the frequency of the sinusoid that best fits the seismic trace. The equation can be further modified as:

$$f_i(t) = \frac{1}{2\pi} \frac{d}{dt} \arctan \left[\frac{y(t)}{x(t)} \right].$$

(2.4)

The instantaneous frequency appears to be noisy in seismic section due to the differentiation that boosts higher frequencies and suppresses lower frequencies. This issue is resolved or the effect is reduced using non-linear filtering or weighting averaging techniques.

2.3.2. Sweetness

The sweetness attribute is used to identify "sweet spots" in the data that are associated with oil and gas deposits. It is defined as the ratio between RMS amplitude and the square root of instantaneous frequency, and is expressed mathematically as:

$$S(t) = \frac{A_{rms}(t)}{\sqrt{f_i(t)}}.$$

(2.5)

It has been demonstrated by Kumar et al. (2019a) that the sweetness anomalies are more obvious than the reflection strength anomalies (Figure 2.4). Hart (2008) documented the significant application of the sweetness attribute in delineating submarine channel belts from seismic data.

2.3.3. Relative Amplitude Change and Instantaneous Bandwidth

The relative amplitude $\sigma_r(t)$ is defined as the rate of change of the instantaneous amplitude $A_i(t)$, and helps in bringing out minute changes in amplitude. This highlights the zone of interference, and is expressed mathematically as:

$$\sigma_r(t) = \frac{\frac{d(A_i(t))}{dt}}{A_i(t)}.$$

(2.6)

Further, the absolute value of this relative amplitude change generates the instantaneous bandwidth $b_i(t)$ (Barnes, 1993), which can be written as

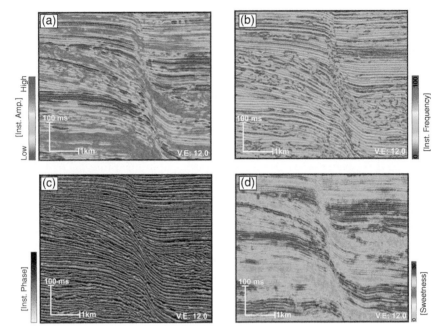

Figure 2.4   (a) Instantaneous amplitude, (b) instantaneous frequency, (c) instantaneous phase, and (d) sweetness attributes computed from seismic data in Taranaki Basin, offshore NZ. Sweetness anomalies are more prominent than the instantaneous amplitude anomalies in defining subsurface (after Kumar et al., 2019d).

$$b(t) = \frac{1}{2\pi} \left| \frac{\frac{d(A_i(t))}{dt}}{A_i(t)} \right|. \tag{2.7}$$

This is used in estimating the spectral bandwidth within the given amplitude change.

2.3.4. RMS Frequency

The RMS frequency (f_{rms}) can be expressed in terms of instantaneous equivalents, defined as the square root of sum of the squares of instantaneous frequency ($f_i(t)$) and instantaneous bandwidth ($b_i(t)$) (Barnes, 1993, 2016). Mathematically these measures can be related as:

$$f_{rms} = \sqrt{f_i(t)^2 + b_i(t)^2}. \tag{2.8}$$

This generates the spectral estimate of f_{rms} at a given instant of time and is always positive or $\geq$ instantaneous frequency. The f_{rms} is a useful measure of the dominant spectral frequency (Papoulis, 1984).

2.3.5. Q-Factor

Quality (Q) factor or instantaneous q-factor, $q_i(t)$, is defined as the ratio between the instantaneous frequency and instantaneous decay rate (Barnes, 1990, 1993; Tonn, 1991). The instantaneous decay rate $\sigma_i(t)$ is defined as the derivative of instantaneous amplitude divided by the instantaneous amplitude (Barnes, 1990, 1991). Mathematically it can be expressed as:

$$q(t) = -\frac{\pi f(t)}{\sigma_i(t)}.$$

(2.9)

The quality factor can thus be considered as a ratio of frequency and amplitude decay or else it can be articulated as the ratio of frequency and bandwidth, which is in agreement with standard definitions (Close, 1966; Johnston and Toksöz, 1981)

Further, the average q-factor is defined as the ratio between the average frequency and twice bandwidth (Barnes, 2016), and is expressed as:

$$q_a(t) = -\frac{\pi f_a(t)}{\sigma_a(t)}.$$

(2.10)

Q-factor is used for distinguishing spectral anomalies in understanding attenuation characteristics of a medium, especially for hydrocarbon exploration and seismological studies.

2.4. Structural and Stratigraphic Attributes

Seismic properties pertaining to the geologic structures and their stratigraphic step-ups are generally characterized by structural and stratigraphic attributes. Structural attributes quantify the structural properties of geologic features, e.g., slope, azimuth, dip, curvature, relief, and discontinuity, whereas stratigraphic attributes in general elaborate the reflection patterns related to the stratigraphy, which includes amplitude variance, reflection spacing, parallelism, and divergence. These attributes and their mathematical formulations are described in the following sections.

2.4.1. Dip and Azimuth Attributes

The dip or dip angle and azimuth attributes define the orientations of reflectors, and are basic 3D structural attributes. These attribute volumes describe the local reflector surface over which interpreters can provide a measure of

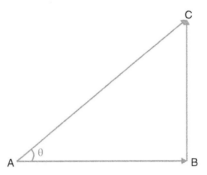

Figure 2.5   Demonstration of dip computation using simplified trigonometric approach. For more clarity on dip measurements please refer to Figure 1.5.

discontinuity (Al-Shuhail et al., 2017). Moreover these attributes, by combining with apparent dip and shaded relief maps, produce significant insights for the interpretation of subsurface discontinuities.

Figure 2.5 shows a simplified approach for dip computation by using a trigonometric function. AB is the horizontal axis and BC is the vertical axis. Hence, dip is the angle that the segment AC makes with the horizontal and is given as:

$$\tan\theta = \frac{BC}{AB}$$

or

$$dip\ angle,\ (\theta_d) = \tan^{-1}\left(\frac{BC}{AB}\right). \tag{2.11}$$

Let us have a look at the dip-azimuth estimates for a 3D environment. Geologically a formation top or a bedding surface can be defined by means of apparent dips θ_x and θ_y or more commonly the true dip θ_d of the surface and its strike ψ_s (Chopra & Marfurt, 2007). Apparent dips θ_x and θ_y are the angular measurements in the vertical (x, z) and (y, z) plane from the horizontal x-axis and y-axis to the interface. These apparent dips, when measured, are also referred as the inline and xline time dips (p and q).

Assuming a constant velocity v, the inline and crossline time dips can be related to their angular dips and can be written as:

$$p = (2\ tan\ \theta_x)/v$$
$$q = (2\ tan\ \theta_y)/v$$

These dips can be combined to obtain the true geological dip θ_d, which is given as:

$$\theta_d = \sqrt{p^2 + q^2}. \tag{2.12}$$

This estimate of the true geological dip is more explicitly called the dip magnitude or the dip angle attribute.

Further, the azimuth (φ) is measured from the y-axis and is given by:

$$(\varphi) = ATAN2\ (q,p). \tag{2.13}$$

The azimuth is generally measured from the north or for convenience from the inline seismic survey axis, which is generally perpendicular to the geological strike, and is measured in the direction of maximum downward dip.

The dip components can be used in computing apparent dip for a certain azimuthal angle (φ). The relationship (Chopra & Marfurt, 2007; Marfurt & Kirlin, 2001) can be expressed as:

$$p\ (\varphi) = p\cos\varphi + q\sin\varphi, \tag{2.14}$$

where $p\ (\varphi)$ is the apparent dip measured at an azimuthal angle φ, and p and q are the respective inline and crossline components.

Several methods for the computation of dip-azimuth volumes include dip scanning, complex trace analysis, plane-wave destructor, and gradient-squared tensor. Applications of dip and azimuth attributes are shown in Figures 2.6 and 2.7, respectively; further applications are available in other literature

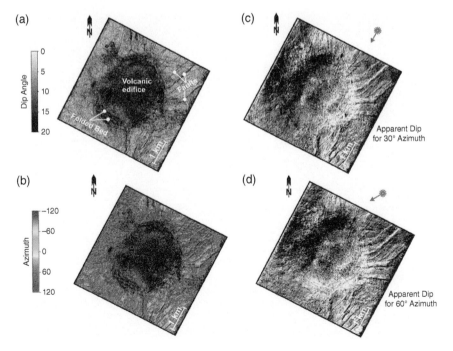

Figure 2.6 (a) Dip angle and (b) azimuth attributes demonstrating the structures of volcanic edifice and discontinuous features; apparent dip attributes for (c) 30° and (d) 60° azimuths displaying subsurface structural elements in depth slices from Kora magmatic prospect in Taranaki Basin, offshore NZ (after Kumar et al., 2019d).

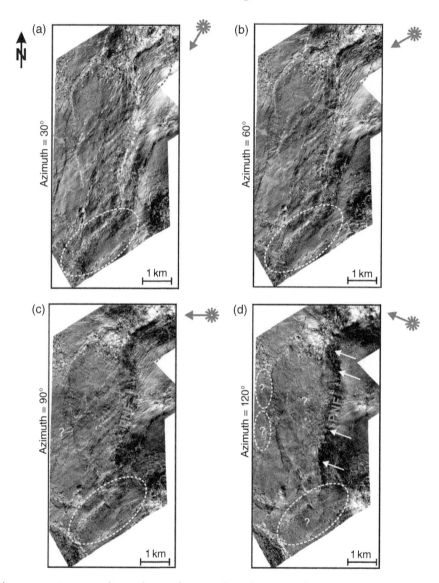

Figure 2.7   Apparent dip attribute in horizon slices for (a) 30°, (b) 60°, (c) 90°, and (d) 120° azimuth showing subsurface structural elements in Parihaka prospect, Taranaki Basin, offshore NZ. The direction of illumination is indicated by red arrow (after Kumar et al., 2019d).

(Gresztenkorn & Marfurt, 1999; Höcker & Fehmers, 2002; Luo et al., 1996, 2006; Marfurt et al., 1998). Readers are directed to these important articles to receive more scientific insights into the dip-azimuths.

Slope and Dip Exaggeration

Exaggeration in reflection slopes and dips improves the display of subsurface structures. This exaggeration factor typically ranges between 5 to 20. The exaggerated dip (e_{dip}) can be related to the exaggerated slope (e_{slope}) through the following equation:

$$e_{dip} = \tan^{-1}(\gamma e_{slope}), \tag{2.15}$$

where γ is the exaggeration factor.

Dip-Steering

Dip-steering is the most modern application for the estimation of dip-azimuth volume, which is also referred as a steering cube or a dip-azimuth volume (Jaglan et al., 2015; Kumar & Mandal, 2017; Kumar & Sain, 2018; Kumar et al., 2019b, c, d; Tingdahl, 1999; Tingdahl & de Groot, 2003). The deep-steering cube contains seismic dip and azimuth information at every sample position. This is calculated by transforming a sub-cube into the 3D Fourier domain and then estimating the maximum dip with the help of a third-order polynomial curve fitting (Tingdahl, 2003) to sub-cube around the sample of the highest energy in the Fourier domain. Then a search for the local maxima is made and the corresponding dip-azimuth to the local maxima is set as the output. This operation, when carried over a 3D cube, generates a dip-azimuth volume. Further, this dip-azimuth volume is smoothed using the structure-oriented filter (SOF) to enhance the subsurface images of geologic structures (Kumar & Mandal, 2017; Kumar & Sain, 2018). For further details on dip-steering, readers are advised to look into Appendix B.

2.4.2. Coherence Attribute

Coherency is a measure of lateral change in seismic responses caused by the variation in structure, stratigraphy, and lithology (Bahorich & Farmer, 1995; Marfurt et al., 1998). Seismic coherency is useful for 3D visualization of discontinuities like faults or stratigraphic features such as the channels, point bars, slumps, tidal drainage patterns, etc. The coherency (Bahorich & Farmer, 1995) is based on the classical normalized cross-correlation technique applied over 3D seismic volume. Let ρ_x be the inline l lag cross-correlation at time t between data traces u at position (x_i, y_i) and (x_{i+1}, y_i). Then ρ_x is given by:

$$\rho_x(t, l, x_i, y_i) = \frac{\sum_{\tau=-w}^{+w} u(t-\tau, x_i, y_i) u(t-\tau-l, x_{i+1}, y_i)}{\sqrt{\sum_{\tau=-w}^{+w} u^2(t-\tau, x_i, y_i) \sum_{\tau=-w}^{+w} u^2(t-\tau-l, x_{i+1}, y_i)}},$$

where $2w$ is the temporal length of the correlation window. Similarly, let ρ_y be the crossline m lag cross-correlation at time t between data traces u at position (x_i, y_i) and (x_i, y_{i+1}). Then ρ_y is given by:

$$\rho_x(t, m, x_i, y_i) = \frac{\sum_{\tau=-w}^{+w} u(t-\tau, x_i, y_i)u(t-\tau-m, x_i, y_{i+1})}{\sqrt{\sum_{\tau=-w}^{+w} u^2(t-\tau, x_i, y_i) \sum_{\tau=-w}^{+w} u^2(t-\tau-m, x_i, y_{i+1})}}.$$

The inline and crossline, l and m lags, are the cross-correlation coefficients that are further combined to obtain a 3D estimate of coherency, defined as ρ_{xy}:

$$\rho_{xy} = \sqrt{[max_{(l)}\rho_x(t, l, x_i, y_i)][max_{(m)}\rho_x(t, m, x_i, y_i)]},$$

where $max_{(l)}\rho_x(t, l, x_i, y_i)$ and $max_{(m)}\rho_x(t, m, x_i, y_i)$ represents the lags l and m for which ρ_x and ρ_y are the maximum.

The cross-correlation algorithm (Bahorich & Farmer, 1995) is based on two orthogonal cross-correlations derived from three traces in an L-shaped pattern. Marfurt et al. (1998) documented that for data, contaminated with coherent noise, such type of cross-correlation algorithm often results in a noisy estimate of coherency. Such three-point cross-correlation algorithm assumes zero-mean seismic signals. Computation of coherency using a larger correlation window intermingles the stratigraphy related to both shallower and deeper times including the zone of interest. Moreover, shortening the window shall further include artefacts that may hinder steady interpretation. A rigorous moving analysis window centered at every point may mitigate these limitations. Hence, Marfurt et al. (1998) developed a semblance-based coherency algorithm and demonstrated that a rectangular analysis window containing J traces centered around the analysis point is very efficient. The semblance $\sigma(\tau, p, q)$, defined for a local planar event at time τ, where p and q are the apparent dips in x and y directions, is given by:

$$\sigma(\tau, p, q) = \frac{\left[\sum_{j=1}^{J} u\left(\tau - px_j - qy_j, x_j, y_j\right)\right]^2 + \left[\sum_{j=1}^{J} u^H\left(\tau - px_j - qy_j, x_i, y_j\right)\right]^2}{J\sum_{j=1}^{J}\left\{\left[u\left(\tau - px_j - qy_j, x_j, x_j\right)\right]^2 + \left[u^H\left(\tau - px_j - qy_j, x_j, y_j\right)\right]^2\right\}},$$

where the superscript H refers to the Hilbert Transform of the real seismic trace u.

This semblance calculation leads to robust estimation. The modified coherency, estimated based on semblance, is given by:

$$C(\tau, p, q) = \frac{\sum_{k=-K}^{+K}\left\{\left[\sum_{j=1}^{J} u\left(\tau + k\Delta t - px_j - qy_j, x_j, y_j\right)\right]^2 + \left[\sum_{j=1}^{J} u^H\left(\tau + k\Delta t - px_j - qy_j, x_j, y_j\right)\right]^2\right\}}{J\sum_{k=-K}^{+K}\sum_{j=1}^{J}\left\{\left[u\left(\tau + k\Delta t - px_j - qy_j, x_j, y_j\right)\right]^2 + \left[u^H\left(\tau + k\Delta t - px_j - qy_j, x_j, y_j\right)\right]^2\right\}},$$

$$(2.16)$$

where K denotes the half-height vertical analysis window and Δt is the temporal sample increment. The output of the coherence measures ranges between 0 and 1, where 0 refers to the minimum trace similarity and 1 refers to the maximum trace similarity. The use of a structural dip improves the performance of the coherence attribute (Marfurt et al., 1999).

2.4.3. Similarity Attribute

Similarity quantifies the similar nature of two segments of seismic traces $u(x, y, t)$ (Tingdahl, 2003; Tingdahl & de Rooij, 2005). The similarity (S) between two trace segments at (x_a, y_a) and (x_b, y_b), centered at time t (Tingdahl, 2003), is expressed as:

$$S_m = 1 - \frac{|a - b|}{|a| + |b|},$$

where:

$$a = \begin{bmatrix} u(x_A, y_A, t + t_1) \\ u(x_A, y_A, t + t_1 + dt) \\ \cdots \\ u(x_A, y_A, t + t_2 - dt) \\ u(x_A, y_A, t + t_2) \end{bmatrix}, b = \begin{bmatrix} u(x_B, y_B, t + t_1) \\ u(x_B, y_B, t + t_1 + dt) \\ \cdots \\ u(x_B, y_B, t + t_2 - dt) \\ u(x_B, y_B, t + t_2) \end{bmatrix},$$

dt is the sampling interval, t_1 is the relative start time of the comparison window, t_2 is the relative stop-time of the comparison window, and u is the amplitude value in seismic cube. When the similarity attribute is computed using dip-steering as an input, the output is called the dip-steered similarity (Tingdahl, 2003), which is expressed as:

$$S_{dip} = 1 - \frac{|a_{dip} - b_{dip}|}{|a_{dip}| + |b_{dip}|}, \tag{2.17}$$

where

$$a_{dip} = \begin{bmatrix} u(x_A, y_A, t_A + t_1) \\ u(x_A, y_A, t_A + t_1 + dt) \\ \cdots \\ u(x_A, y_A, t_A + t_2 - dt) \\ u(x_A, y_A, t_A + t_2) \end{bmatrix}, b_{dip} = \begin{bmatrix} u(x_B, y_B, t_B + t_1) \\ u(x_B, y_B, t_B + t_1 + dt) \\ \cdots \\ u(x_B, y_B, t_B + t_2 - dt) \\ u(x_B, y_B, t_B + t_2) \end{bmatrix},$$

t_A and t_B are the dip-steered times going from the position (x, y, t) to the traces at (x_A, y_A) and (x_B, y_B), respectively, and u is the amplitude value in the seismic cube. Application of amplitude and similarity attributes for discerning faults is shown in Figure 2.8.

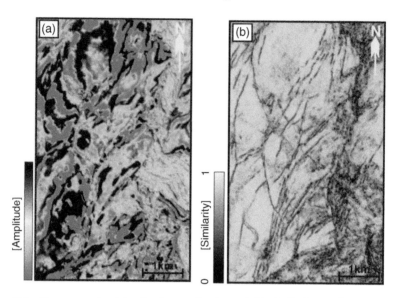

Figure 2.8   Time map for (a) seismic amplitude and (b) similarity attribute displaying geologic faults and associated discontinuities over Parihaka prospect, Taranaki Basin, offshore NZ (after Kumar et al., 2019d).

2.4.4. Curvature Attribute

Curvature is a measure of how much a surface is bent at a particular point. The more bent the surface is, the greater its curvature (Chopra & Marfurt, 2007; Roberts, 2001). For a particular point on a curve, its curvature is defined as the rate of change of direction of the curve (Figure 2.9). Two-dimensionally, the curvature is defined as the inverse of the radius of the osculating circle at a given point on a curve (Roberts, 2001; Sigismondi & Soldo, 2003) and can be mathematically expressed as:

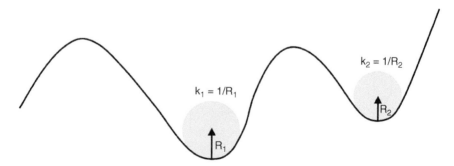

Figure 2.9   Curvature defined over a curved surface.

$$k_{2d} = \frac{d\omega}{ds} = \frac{d^2y/dx^2}{\left(1 + (dy/dx)^2\right)^{3/2}} = \frac{1}{R}, \tag{2.18}$$

where $d\omega$ is the rate of change of angle with respect to the arc length ds, and R is the radius of curvature.

For a three-dimensional case, this idea can be extended by imagining a surface being intersected by an orthogonal set of two vertical planes (Roberts, 2001; Sigismondi and Soldo, 2003). The intersection of the orthogonal planes produces curves on the surface, for which curvature can be computed by fitting a quadratic surface to the chosen horizon from the formation top, in a least squares sense. However, such estimation becomes erroneous when the chosen horizon contains no continuous reflector. It should be noted that such situations are common when seismic reflectors are associated with faults and discontinuous features. The computation of volumetric curvature (de Rooij & Tingdahl, 2002) alleviates such problems where the calculation is carried out using the dip-steering data volume.

A quadratic surface that fits to the chosen horizon can be written as:

$$z(x,y) = ax^2 + cxy + by^2 + dx + ey + f. \tag{2.19}$$

The dip components can be obtained from the dip-steering data volume. These dip components can be further related to the z function as:

$p = \partial z/\partial x$, where p is the inline dip component
$q = \partial z/\partial y$, where q is the xline dip component.

Once these components are obtained, the coefficients of the z function are determined as:

$$a = \frac{1}{2}\frac{\partial p}{\partial x} = \frac{1}{2}\frac{\partial}{\partial x}\frac{\partial z}{\partial x} \tag{2.20}$$

$$b = \frac{1}{2}\frac{\partial q}{\partial y} = \frac{1}{2}\frac{\partial}{\partial y}\frac{\partial z}{\partial y} \tag{2.21}$$

$$c = \frac{1}{2}\left[\frac{\partial q}{\partial x} + \frac{\partial p}{\partial y}\right] = \frac{1}{2}\left[\frac{\partial}{\partial x}\frac{\partial z}{\partial y} + \frac{\partial}{\partial y}\frac{\partial z}{\partial x}\right] \tag{2.22}$$

$$d = p \tag{2.23}$$

$$e = q. \tag{2.24}$$

This set of coefficients can now be used to yield various curvature measures. The mean and Gaussian curvatures can be mathematically expressed as:

$$k_{mean} = \frac{a\left(1 + e^2\right) + b\left(1 + d^2\right) - cde}{\sqrt[3]{\left(1 + d^2 + e^2\right)}} \tag{2.25}$$

or

$$k_{mean} = \frac{\frac{1}{2}\frac{\partial p}{\partial x}(1 + e^2) + \frac{1}{2}\frac{\partial q}{\partial y}(1 + p^2) - \frac{1}{2}pq\left(\frac{\partial q}{\partial x} + \frac{\partial p}{\partial y}\right)}{\sqrt[3]{(1 + p^2 + q^2)}} \quad (2.26)$$

$$k_{gauss} = \frac{4ab - c^2}{\left(1 + d^2 + e^2\right)^2} \quad (2.27)$$

or

$$k_{gauss} = \frac{2a \times 2b - c^2}{\left(1 + d^2 + e^2\right)^2} \quad (2.28)$$

or

$$k_{gauss} = \frac{\frac{\partial p}{\partial x}\frac{\partial q}{\partial y} - \frac{1}{4}\left[\frac{\partial^2 q}{\partial x^2} + \frac{\partial^2 p}{\partial y^2} + 2\frac{\partial q}{\partial x}\frac{\partial p}{\partial y}\right]}{\left(1 + p^2 + q^2\right)^2}. \quad (2.29)$$

The mean and Gaussian curvatures can then be used to derive the maximum and minimum curvatures. The mathematical formulations pertaining to these curvatures are given by:

$$k_{max} = k_{mean} + \sqrt{k_{mean}^2 - k_{gauss}} \quad (2.30)$$

$$k_{min} = k_{mean} - \sqrt{k_{mean}^2 - k_{gauss}}. \quad (2.31)$$

The mean, gauss, maximum, and minimum curvatures are related as:

$$k_{mean} = \frac{k_{max} + k_{min}}{2} \quad (2.32)$$

and

$$k_{gauss} = k_{max} \times k_{min}. \quad (2.33)$$

The most positive curvature k_{pos} is mathematically expressed as:

$$k_{pos} = (a + b) + \left[(a - b)^2 + c^2\right]^{1/2} \quad (2.34)$$

or

$$k_{pos} = \frac{1}{2}\left[\frac{\partial p}{\partial x} + \frac{\partial q}{\partial y}\right] + \left[\frac{1}{4}\left(\frac{\partial p}{\partial x} - \frac{\partial q}{\partial y}\right)^2 + \frac{1}{4}\left(\frac{\partial q}{\partial x} + \frac{\partial p}{\partial y}\right)^2\right]^{\frac{1}{2}}. \quad (2.35)$$

Similarly, the most negative curvature k_{neg} is mathematically expressed as:

$$k_{neg} = (a + b) - \left[(a - b)^2 + c^2\right]^{1/2} \tag{2.36}$$

or

$$k_{neg} = \frac{1}{2}\left[\frac{\partial p}{\partial x} + \frac{\partial q}{\partial y}\right] - \left[\frac{1}{4}\left(\frac{\partial p}{\partial x} - \frac{\partial q}{\partial y}\right)^2 + \frac{1}{4}\left(\frac{\partial q}{\partial x} + \frac{\partial p}{\partial y}\right)^2\right]^{\frac{1}{2}}. \tag{2.37}$$

Apart from these curvature properties, the coefficients described in Eqs. 2.19 to 2.24 can also be used to derive several other curvatures, e.g., dip curvature, strike curvature, shape index, etc. The most widely used curvatures are the maximum, minimum, most positive, and most negative curvatures. Readers are advised to look at Chapter 3 for the application and usage of the curvature attributes.

2.4.5. Advanced Structural Attributes

Ridge Enhancement Filter (REF) Attribute
The REF attribute was proposed by Brouwer & Huck (2011), which takes the similarity attribute as an input to improve fault visibility by sharpening the edges and maintaining the connectivity of fault networks. It is computed by extracting nine similarity values surrounding the central evaluation point, and is done by scanning the target in four different directions (Figure 2.10a), i.e. the inline and the xline and two diagonals oriented at 45° and 135° in a time slice domain of the similarity

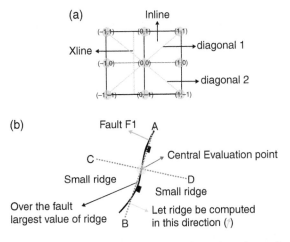

Figure 2.10 (a) Demonstration of REF computation along the inline, xline, and diagonal directions. (b) Different values of ridge observed along the fault. The illustrations in (a) and (b) show the analysis involved in REF calculation.

attribute volume. Let us consider the Fault F1 (shown in Figure 2.10b), where AB and CD are two segments crossing the fault. The central evaluation point is shown through a blue transparent circle. Thus, the ridge is defined by the following equation:

$$Ridge = \frac{Sum\ of\ the\ values\ on\ either\ side\ of\ the\ fault}{2} - the\ central\ value. \tag{2.38}$$

Either side of the fault will output small values of the ridge, whereas the places over the fault are associated with large ridge values. The application of REF has been demonstrated in Chapters 8 and 9.

Thin Fault Likelihood (TFL) Attribute

The TFL attribute aims to output thinned sharp fault images from seismic data (Hale, 2013). The computation uses the principle of fault-oriented semblance. Using the reflection slopes, initially a structure-oriented semblance is computed by the following expression as:

$$semblance = \frac{\langle image \rangle_s^2}{\langle image^2 \rangle_s}, \tag{2.39}$$

where a structure-oriented averaging $\langle \bullet \rangle_s$ operation is performed. Such computation results in large semblance values for small numerators and denominators. These variations could be reduced by additionally smoothing the numerators and denominators. By incorporating this additional smoothing, the semblance ratio can be defined as:

$$smooth\ (semblance) = \frac{\left\langle \langle image \rangle_s^2 \right\rangle_f}{\left\langle \langle image^2 \rangle_s \right\rangle_f}, \tag{2.40}$$

where $\langle \bullet \rangle_f$ denotes additional smoothing.

After applying such smoothing, the fault likelihood attribute (f_l) is then defined as:

$$f_l \equiv 1 - (semblance)^8. \tag{2.41}$$

The power term used this equation helps to discriminate between the samples corresponding to larger and smaller likelihood of fault. The computation of f_l becomes accurate if prior information on fault orientation and location is available. Scanning the numerator and denominator in the direction of fault strikes and dips gives accurate alignment of structures in their respective orientations and improves structural imagery from the data (Hale, 2013). The TFL output values range between 0 and 1, where 0 refers to the minimum fault likelihood and 1 refers to the maximum likelihood of faults. The similarity, coherence, most

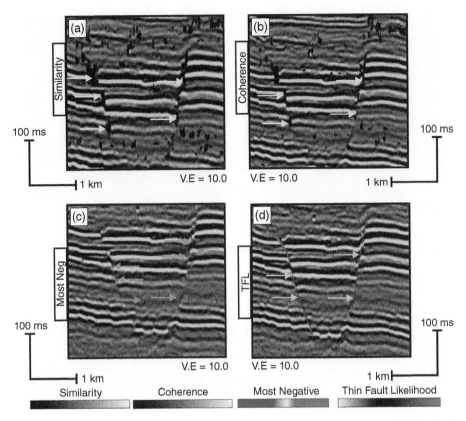

Figure 2.11   (a) Similarity, (b) coherency, (c) most negative curvature, and (d) thin fault likelihood attributes, exhibiting interpretation of structural discontinuities from seismic data in the Taranaki Basin, offshore NZ (after Kumar et al., 2019d). The fault zones are associated with low similarity and coherency. The downthrown portion of the faults is captured by the negative curvature. The thin fault likelihood attribute brings out thinned razor-shaped fault images. The structural discontinuities observed from each of the attributes (a–d) are highlighted using yellow, orange, and pink arrows.

negative curvature, and TFL attributes are shown in Figure 2.11. The readers may refer to Kumar & Sain (2018) and Kumar et al. (2019d) for a detailed description of TFL and its application.

Pseudo Relief Attribute

The pseudo relief seismic attribute, consistent with seismic images of geologic structures, provides the relief of subsurface structures from seismic data. The attribute is obtained initially by computing the RMS amplitude $\overline{(A_{rms})}$ from seismic data volume within a defined window, which is mathematically expressed as:

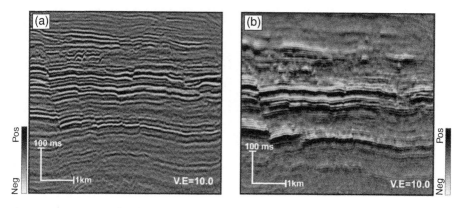

Figure 2.12 (a) Amplitude section and (b) corresponding pseudo relief attribute, having improved structural visibility that aids in subsurface interpretation of geologic features (After Kumar & Sain, 2018; Kumar et al., 2019d).

$$\overline{A_{rms}} = \sqrt{\frac{1}{N} \sum_{i=1}^{N} a_i^2}, \qquad (2.42)$$

where a_i corresponds to the trace amplitude and N is the number of samples. Once $\overline{A_{rms}}$ is computed, a phase of $-90°$ is rotated through inverse Hilbert Transform (HT), which finally produces the pseudo relief (r_{pseudo}) attribute, which is expressed as:

$$r_{pseudo} = H^{-1}\left\{ \overline{A_{rms}} \right\}, \qquad (2.43)$$

where H^{-1} is the inverse Hilbert Transform (Appendix A).

The HT operator is an ideal phase rotator and results in obtaining a change of 90° in the signal phase. The phase rotation operation aims to convert the positive values of amplitudes after RMS calculation into both positive and negative values (Purves, 2014; Vernengo & Trinchero, 2015). The output of this attribute is demonstrated in Figure 2.12. More descriptions of the application of this attribute are available in Chapters 8 and 9.

2.4.6. Amplitude Variance

The amplitude variance attribute is useful in highlighting subsurface stratigraphy by illuminating the variation in reflection amplitude. It is defined as the variance in amplitude normalized over the square of the average amplitudes. The input attribute used for its computation is generally an instantaneous amplitude

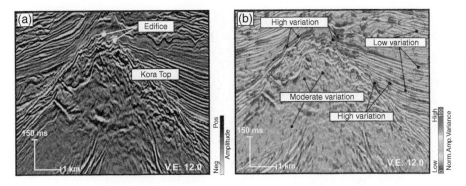

Figure 2.13   (a) Amplitude section and (b) corresponding normalized amplitude variance attribute displaying variation in amplitude that can be used for interpretation of high-, moderate-, and low-energy depositional environment.

or the reflection strength obtained from the seismic cube within a defined window. Mathematically the amplitude variance is expressed as:

$$V_n = \frac{\sigma_{amp}^2}{\mu_{amp}^2},\qquad(2.44)$$

where V_n is the normalized amplitude variance, σ_{amp} is the standard deviation of the amplitudes, and μ_{amp} is the average of the amplitudes. The V_n can further be scaled by taking its square root and then multiplying by 100. This is expressed as:

$$V_n\ (scaled) = 100\sqrt{V_n}.\qquad(2.45)$$

This scaling outputs amplitude variance within a range between 0 to 100. This range can be used to segregate the attribute response into uniform amplitudes, moderately varying amplitudes, and highly varying amplitudes. Based on the strength, the normalized amplitude variance attribute can decipher the high-, medium-, and low-energy depositional environment (Figure 2.13).

2.4.7. Reflection Spacing

The reflection spacing attribute (Figure 2.14) provides a qualitative measure, and is defined as the distance between successive reflections measured perpendicular to reflections. The attribute response is very similar to that of the RMS frequency.

2.4.8. Reflection Divergence

The reflection divergence attribute measures the amount by which reflections spherically diverge from a given sequence. Towards the base of such a sequence, reflections converge and pinch-outs are observed. This attribute mainly suggests

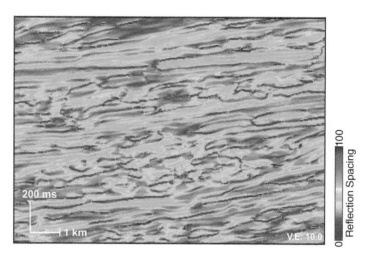

Figure 2.14   Reflection spacing, an instantaneous attribute displaying the arrangement of reflections within different stratigraphic horizons.

the presence of a high depositional prograding environment. Empirically, it is quantified as:

$$R(divergence) = c\left\langle \frac{\partial p}{\partial t} \right\rangle . W(\phi), \qquad (2.46)$$

where p is the reflection slope, c is a constant that scales the divergence attribute, and W is a weighting function of azimuth ϕ.

2.4.9. Reflection Parallelism

The reflection parallelism attribute defines the variations in reflection orientations (whether parallel or non-parallel) within a given stratigraphic sequence. Parallel reflections that possess uniform dip-azimuth and non-parallel reflections that exhibit variable dip and azimuth are associated with low-energy to high-energy depositional environments (de Rooij & Tingdahl, 2002). The output of the reflection parallelism can be conveniently tuned to a scale ranging between 0 to 100 (Eqs. 2.46 and 2.47 and Figure 2.15), where the larger values are associated with parallel reflections and lower values signify non-parallel reflections.

$$R(Spacing) = \overline{Norm\ Var(\theta, \phi)} \qquad (2.47)$$

$$R_{normalized}(Spacing) = 100(R(spacing), \qquad (2.48)$$

where θ and ϕ are the dip and azimuth of the reflections.

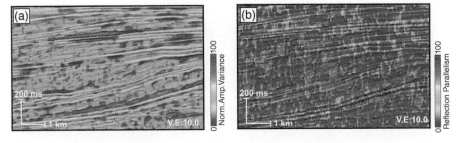

Figure 2.15 (a) Computed normalized amplitude variance and (b) reflection parallelism attributes displaying reflection patterns within a depositional environment.

2.4.10. Spectral Decomposition

The spectral decomposition (SD) technique decomposes seismic signals into different frequency sub-bands. This is an important approach to visualize a feature of interest in the frequency domain rather than in the time domain. The SD method has been fruitful in a number of applications, e.g., channel detection, stratigraphic visualization, reservoir characterization, direct indication of hydrocarbons, etc. (Castagna et al., 2003; Li & Zheng, 2008; Liu et al., 2017; Marfurt & Kirlin, 2001; Sinha et al., 2005; Sun et al., 2002). The decomposition algorithms, i.e. the Short-Time Fourier Transform (STFT) or Continuous Wavelet Transform (CWT), have proven their worth for delineation and characterization of subsurface features from seismic data (Figure 2.16), as has been described by Kumar et al. (2021).

2.4.11. Velocity, Reflectivity, and Attenuation Attributes

Seismic velocity, reflectivity, and attenuation attributes have been used for the delineation of gas hydrates—future major potential energy resources. The presence of gas hydrates is mainly detected by identifying an anomalous reflector, known as the bottom simulating reflector or BSR on seismic section. The BSR often lies at the base of the gas hydrates stability zone (Sain & Gupta, 2008, 2012; Sain et al., 2011), and is not a geological interface but a physical reflector that is caused by high-velocity hydrates-bearing sediments above and low-velocity gas-bearing sediments below. Gas hydrates are characterized by several attributes, such as high seismic velocity (Figure 2.17) and low attenuation (Q^{-1}) (Figure 2.18) attributes, whereas underlying gas-saturated sediments are associated with low velocity (Figure 2.17) and high reflectivity (Figure 2.19) attributes (Jyothi et al., 2017; Kumar et al., 2019a; Ojha & Sain, 2009; Ramu and Sain, 2019, 2021; Ramu et al., 2021; Sain & Singh, 2011; Sain et al., 2009; Satyavani et al., 2015, 2017). The

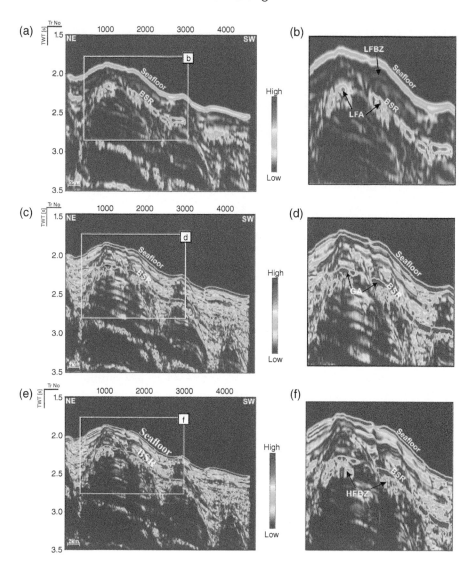

Figure 2.16 Comparison of CWT spectral responses using Mexican Hat mother wavelet for (a–b) 15 Hz, (c–d) 35 Hz, (e–f) 60 Hz frequencies respectively in Mahanadi Basin, eastern Indian offshore. The right panel is the zoomed version of the box shown in the left panel. As better resolved, the 35 Hz frequency is considered to be the tuning frequency for said data set. LFBZ: Low Frequency Blackout Zone; HFBZ: High-Frequency Blackout Zone (after Kumar et al., 2021).

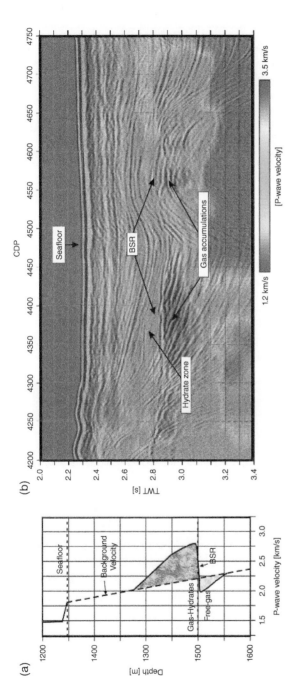

Figure 2.17 (a) Typical "velocity increase" due to gas hydrates and "velocity drop" by free gas across the BSR against the background (without gas hydrates and free gas) trend. (b) Seismic velocity anomaly showing the lateral and vertical distribution of gas hydrates (high-velocity) and free gas (low-velocity) bearing sediments (after Ojha & Sain, 2009). Source: Ojha & Sain (2009), Figure 06 [p.268]/CC BY-NC 4.0.

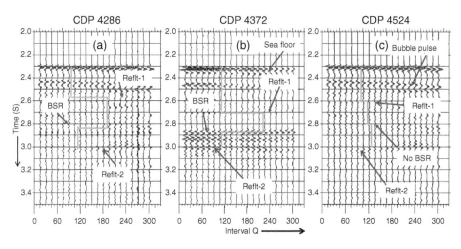

Figure 2.18   High seismic quality factor (Q) or low attenuation above the BSR, showing the presence of gas hydrates at two locations (a, b). This has been compared with no-hydrates at one location (c) (after Sain & Singh, 2011). Blue arrows indicate the interface of seafloor, BSR, and reflectors 1 and 2.

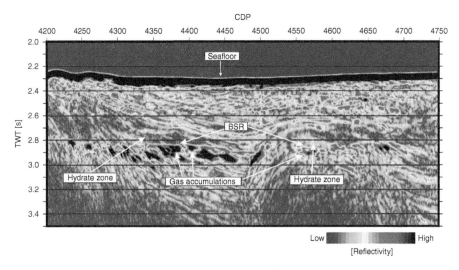

Figure 2.19   High reflectivity anomaly below the BSR, showing the lateral and vertical variation of free gas into the sediments. Source: Ojha & Sain (2009), Figure 04[p.267]/ CC BY-NC 4.0.

BSR-like feature is also developed by the diagenesis. Thus, the computation of these attributes ascertain if the BSR is related to gas hydrates.

2.5. A Glance at Interpretation Pitfalls

Subsurface interpretations may be direct or simple in migrated sections prepared from the data acquired over areas devoid of complex geology. However, analysis of the data acquired from highly structured terrain is associated with several uncertainties that may lead to erroneous or false interpretations (Figures 2.19 and 2.20). These pitfalls are generally produced due to several effects, e.g., geometric, shallow, deep, acquisition, or processing (Tucker & Yorston, 1973).

Geometric effects result in wider convex-up structures, e.g., domes and anticlines and convex-down structures. A bowtie is a common pitfall generally observed over unmigrated seismic sections. A proper migration algorithm should be adopted to escape such pitfalls. Shallow effects are produced to near-surface lateral velocity or thickness changes. This results in false highs in near-surface zones within the data. Thus, a critical quality check is recommended to alleviate these issues. Deep effects are generated due to velocity changes within subsurface strata caused by complex structural/stratigraphic features. For example, the downthrown side of normal faults may be associated with thinned reflections, which happens due to an increase in velocity with depth. False bowl shape

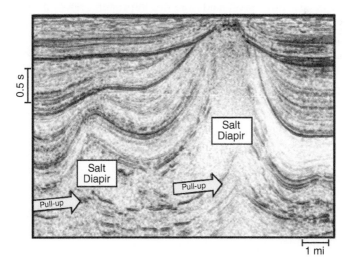

Figure 2.20 Seismic section in the southern North Sea demonstrating pull-up artefact created by higher velocity of Zechstein salt diapir in the interior as compared to the flanks, leading to false geologic interpretation. Source: Marfurt & Alves (2015), Figure 03 [p.A09]/CC BY 4.0.

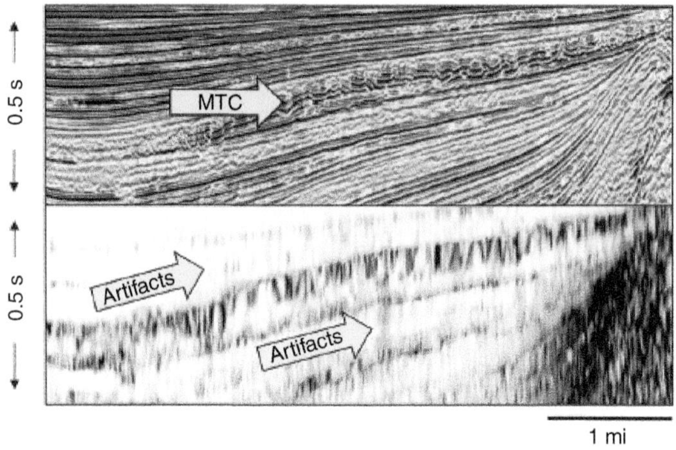

Figure 2.21   Coherency profile from SE Brazil showing false vertical fabrics in Miocene MTC. MTC: Mass Transport Complex. Source: Marfurt & Alves (2015), Figure 06[p.A12]/ CC BY 4.0.

features may be associated at deeper levels, which are generally referred to as migration smiles. Thus, a validity check is done to inspect whether these anomalies are related to geologic structures or are simply artefacts. Acquisition/processing artefacts are also produced due to improper selection of acquisition/ processing parameters. A common pitfall known as acquisition footprints are observed in a cross-section and map view (Figure 2.21 and Figure 2.22). These footprints appear as linear features, which may be misinterpreted for the presence of geologic structures. Thus, a proper quality check and the employment of footprint removal filters are essential to overcome these problems and provide an interpretation of actual geologic features.

2.6. Summary

This chapter has presented definitions and mathematical formulations of different complex trace, structural, and stratigraphic attributes that play a pivotal role in the interpretation of subsurface structures and stratigraphy from seismic data. Moreover, this chapter cautions the readers about the pitfalls encountered during an interpretation. Complex trace analysis helps to extract the amplitude and phase information that can be optimally used for meaningful interpretation of subsurface geology. Further, the structural attributes such as the dip, azimuth, and curvature, and the stratigraphic attributes like amplitude variance, reflection divergence, parallelism, etc., can be integrated for better delineation of the subsurface features and the depositional environment.

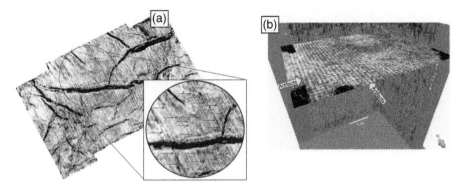

Figure 2.22   (a) Time slice from the Wyandot Formation Top, Penobscot prospect, offshore Nova Scotia, demonstrating footprints of linear features (marked with red arrows in the zoomed view) (adapted from Mandal & Srivastava, 2018). Source: Mandal & Srivastav (2018), Figure 07 [p.42]/with permission of Elsevier. (b) Shallower time slice (at t = 0.40 s) showing the footprints of linear features. Source: Marfurt & Alves (2015), Figure 01 [p.A07]/CC BY 4.0.

The next chapter provides some tasks for simple geologic interpretation from an example section to get acquainted with seismic attributes and geologic features, and to help readers to become practitioners.

References

Al-Shuhail, A. A., Al-Dossary, S. A., & Mousa, W. A. (2017). *Seismic data interpretation using digital image processing*. *John Wiley & Sons*. https://doi.org/10.1002/9781119125594

Barnes, A. E. (1990). Analysis of temporal variations in average frequency and amplitude of COCORP deep seismic reflection data. In *SEG Technical Program Expanded Abstracts* (pp. 1553–1556). SEG. https://doi.org/10.1190/1.1890055

Barnes, A. E. (1991). Instantaneous frequency and amplitude at the envelope peak of a constant-phase wavelet. *Geophysics, 56*(7), 1058–1060. https://doi.org/10.1190/1.1443115

Barnes, A. E. (1993). Instantaneous spectral bandwidth and dominant frequency with applications to seismic reflection data. *Geophysics, 58*(3), 419–428. https://doi.org/10.1190/1.1443425

Barnes, A. E. (Ed.). (2016). *Handbook of poststack seismic attributes*. Society of Exploration Geophysicists. https://doi.org/10.1190/1.9781560803324

Bahorich, M., & Farmer, S. (1995). 3-D seismic discontinuity for faults and stratigraphic features: The coherence cube. *The Leading Edge, 14*(10), 1053–1058. https://doi.org/10.1190/1.1437077

Chopra, S., & Marfurt, K. J. (2007). *Seismic attributes for prospect identification and reservoir characterization*. SEG, Tulsa Society of Exploration Geophysicists and European Association of Geoscientists and Engineers. https://doi.org/10.1190/1.9781560801900

Close, C. M. (1966). *The analysis of linear circuits*. Harcourt, Brace & World, Inc.

Barnes, A. E. (Ed.). (2016). *Handbook of poststack seismic attributes*. Society of Exploration Geophysicists. https://doi.org/10.1190/1.9781560803324

Brouwer, F. & Huck, A. (2011). An integrated workflow to optimize discontinuity attributes for the imaging of faults. In *Attributes: New Views on Seismic Imaging—Their Use in Exploration and Production. GCSSEPM, 31st Annual Conference Publication* (Vol. *496*, p. 533). https://doi.org/10.5724/gcs.11.31.0496

Castagna, J. P., Sun, S., & Siegfried, R. W. (2003). Instantaneous spectral analysis: Detection of low-frequency shadows associated with hydrocarbons. *The Leading Edge, 22*(2), 120–127. https://doi.org/10.1190/1.1559038

de Rooij, M. & Tingdahl, K. M. (2002). Meta-attributes—the key to multivolume, multi attribute interpretation. *The Leading Edge, 21*(10), 1050–1053. https://doi.org/10.1190/1.1518445

Gersztenkorn, A. & Marfurt, K. J. (1999). Eigen structure-based coherence computations as an aid to 3-D structural and stratigraphic mapping. *Geophysics, 64*, 1468–1479. https://doi.org/10.1190/1.1826633

Hale, D. (2013). Methods to compute fault images, extract fault surfaces, and estimate fault throws from 3D seismic images. *Geophysics, 78*(2), O33–O43. https://doi.org/10.1190/geo2012-0331.1

Hart, B. S. (2008). Channel detection in 3-D seismic data using sweetness. *AAPG Bulletin, 92*(6), 733–742. https://doi.org/10.1306/02050807127

Höcker, C., & Fehmers, G. (2002). Fast structural interpretation with structure-oriented filtering. *The Leading Edge, 21*(3), 238–243. https://doi.org/10.1190/1.1598121

Jaglan, H., Qayyum, F., & Hélène, H. (2015). Unconventional seismic attributes for fracture characterization. *First Break, 33*(3), 101–109. https://doi.org/10.3997/1365-2397.33.3.79520

Johnston, D. H. & Toksöz, M. N. (1981). Definitions and terminology. In M. N. Toksöz and D. H. Johnston (Eds.), *Seismic wave attenuation: SEG Geophysics Reprint Series No. 2* (pp. 1–5).

Jyothi, V., Sain, K., Pandey, V., & Bhaumik, A. K. (2017). Seismic attenuation for characterization of gas hydrate reservoir in Krishna-Godavari Basin, eastern Indian margin. *Journal of the Geological Society of India, 90*(3), 261–266. https://doi.org/10.1007/s12594-017-0713-9

Kumar, J., Sain, K., & Arun, K. P. (2019a). Seismic attributes for characterizing gas hydrates: a study from the Mahanadi offshore, India. *Marine Geophysical Research, 40*(1), 73–86. https://doi.org/10.1007/s11001-018-9357-4

Kumar P.C. & Mandal, A. (2017). Enhancement of fault interpretation using multi-attribute analysis and artificial neural network (ANN) approach: A case study from Taranaki Basin, New Zealand. *Exploration Geophysics, 49*(3), 409–424. https://doi.org/10.1071/EG16072

Kumar P.C. & Sain, K. (2018). Attribute amalgamation-aiding interpretation of faults from seismic data: An example from Waitara 3D prospect in Taranaki basin off New Zealand. *Journal of Applied Geophysics, 159*, 52–68. https://doi.org/10.1016/j.jappgeo.2018.07.023

Kumar, P.C., Omosanya, K.O & Sain, K. (2019b). Sill Cube: An automated approach for the interpretation of magmatic sill complexes on seismic reflection data. *Marine and Petroleum Geology, 100*, 60–84. https://doi.org/10.1016/j.marpetgeo.2018.10.054

Kumar, P.C. Sain, K. & Mandal, A. (2019c). Delineation of a buried volcanic system in Kora prospect off New Zealand using artificial neural networks and its implications. *Journal of Applied Geophysics*, *161*, 56–75. https://doi.org/10.1016/j.jappgeo.2018.12.008

Kumar, P.C. Omosanya, K.O, Alves, T. M. & Sain, K. (2019d). A neural network approach for elucidating fluid leakage along hard-linked normal faults. *Marine and Petroleum Geology*, *110*, 518–538. https://doi.org/10.1016/j.marpetgeo.2019.07.042

Kumar, J., Sain, K., & Arun, K.P. (2021). Time-frequency analysis for delineating gas hydrates and free gas in the Mahanadi offshore, India, *Exploration Geophysics*, Published online, https://doi.org/10.1080/08123985.2021.1889365.

Li, Y., & Zheng, X. (2008). Spectral decomposition using Wigner-Ville distribution with applications to carbonate reservoir characterization. *The Leading Edge*, *27*(8), 1050–1057. https://doi.org/10.1190/1.2967559

Liu, W., Cao, S., Wang, Z., Kong, X., & Chen, Y. (2017). Spectral decomposition for hydrocarbon detection based on VMD and Teager–Kaiser energy. *IEEE Geoscience and Remote Sensing Letters*, *14*(4), 539–543. https://doi.org/10.1109/LGRS.2017.2656158

Luo, Y., Higgs, W. G., & Kowalik, W. S. (1996). Edge detection and stratigraphic analysis using 3D seismic data. Paper presented in SEG Technical Program Expanded Abstracts 1996 (pp. 324–327). Society of Exploration Geophysicists. https://doi.org/10.1190/1.1826632

Luo, Y., Wang, Y. E., AlBinHassan, N. M., & Alfaraj, M. N. (2006). Computation of dips and azimuths with weighted structural tensor approach. *Geophysics*, *71*(5), V119-V121. https://doi.org/10.1190/1.2235591

Mandal, A., & Srivastava, E. (2018). Enhanced structural interpretation from 3D seismic data using hybrid attributes: New insights into fault visualization and displacement in Cretaceous formations of the Scotian Basin, offshore Nova Scotia. *Marine and Petroleum Geology*, *89*, 464–478. https://doi.org/10.1016/j.marpetgeo.2017.10.013

Marfurt, K. J., Kirlin, R. L., Farmer, S. L., & Bahorich, M. S. (1998). 3-D seismic attributes using a semblance-based coherency algorithm. *Geophysics*, *63*(4), 1150–1165. https://doi.org/10.1190/1.1444415

Marfurt, K. J., Sudhaker, V., Gersztenkorn, A., Crawford, K. D., & Nissen, S. E. (1999). Coherency calculations in the presence of structural dip. *Geophysics*, *64*(1), 104–111. https://doi.org/10.1190/1.1444508

Marfurt, K. J., & Kirlin, R. L. (2001). Narrow-band spectral analysis and thin-bed tuning. *Geophysics*, *66*(4), 1274–1283. https://doi.org/10.1190/1.1487075

Marfurt, K. J., & Alves, T. M. (2015). Pitfalls and limitations in seismic attribute interpretation of tectonic features. *Interpretation*, *3*(1), SB5–SB15. https://doi.org/10.1190/INT-2014-0122.1

Ojha, M., & Sain, K. (2009). Seismic attributes for identifying gas-hydrates and free-gas zones: application to the Makran accretionary prism. *Episodes*, *32*(4), 264–270. https://doi.org/10.18814/epiiugs/2009/v32i4/003

Papoulis, A. (1984). *Probability, random variables and stochastic processes*. McGraw Hill Book & Co.

Purves, S. (2014). Phase and the Hilbert transform. *The Leading Edge*, *33*, 1164–1166. https://doi.org/10.1190/tle33101164.1

Ramu, R., & Sain, K. (2019). Characterization of Gas Hydrates Reservoirs in Krishna-Godavari Basin, Eastern Indian Margin. *Journal of the Geological Society of India*, *93*(5), 539–545. https://doi.org/10.1007/s12594-019-1215-8

Ramu, R., & Sain, K. (2021). Multi-attribute and artificial neural network analysis of seismic inferred chimney-like features in marine sediments: A study from KG Basin, India. *Journal of the Geological Society of India*, *97*, 238–242. https://doi.org/10.1007/s12594-021-1672-8

Ramu, C., Sunkara, S.L., Ramu, R. & Sain, K. (2021). An ANN-based identification of geological features using multi-attributes: a case study from Krishna-Godavari basin, India, *Arabian Journal of Geosciences*, *14*(4), 1–10. https://doi.org/10.1007/s12517-021-06652-z

Roberts, A. (2001). Curvature attributes and their application to 3 D interpreted horizons. *First Break*, *19*(2), 85–100. https://doi.org/10.1046/j.0263-5046.2001.00142.x

Sain, K. & Gupta, H.K. (2008). Gas hydrates: Indian scenario, *Journal of Geological Society of India*, *72*, 299–311

Sain, K., Singh, A. K., Thakur, N. K., & Khanna, R. (2009). Seismic quality factor observations for gas-hydrate-bearing sediments on the western margin of India. *Marine Geophysical Researches*, *30*(3), 137. https://doi.org/10.1007/s11001-009-9073-1

Sain, K., Rajesh, V., Satyavani, N., Subbarao, K. V., & Subrahmanyam, C. (2011). Gas-hydrate stability thickness map along the Indian continental margin. *Marine and Petroleum Geology*, *28*(10), 1779–1786.

Sain, K., & Singh, A. K. (2011). Seismic quality factors across a bottom simulating reflector in the Makran Accretionary Prism, Arabian Sea. *Marine and Petroleum Geology*, *28*(10), 1838–1843. https://doi.org/10.1016/j.marpetgeo.2011.03.008

Sain, K., & Gupta, H. (2012). Gas hydrates in India: Potential and development. *Gondwana Research*, *22*(2), 645–657. https://doi.org/10.1016/j.marpetgeo.2011.03.013

Satyavani, N., Alekhya, G., & Sain, K. (2015). Free gas/gas hydrate inference in Krishna–Godavari basin using seismic and well log data. *Journal of Natural Gas Science and Engineering*, *25*, 317–324. https://doi.org/10.1016/j.jngse.2015.05.010

Satyavani, N, Sain, K., & Nara, D. (2017). Seismic vis-a-vis sonic attenuation in gas hydrate bearing sediments of Krishna-Godavari basin, Eastern Margin of India. *Geophysical Journal International*, *209*(2), 1195–1203. https://doi.org/10.1093/gji/ggx089

Sigismondi, M. E., & Soldo, J. C. (2003). Curvature attributes and seismic interpretation: Case studies from Argentina basins. *The Leading Edge*, *22*(11), 1122–1126. https://doi.org/10.1190/1.1634916

Sinha, S., Routh, P. S., Anno, P. D., & Castagna, J. P. (2005). Spectral decomposition of seismic data with continuous-wavelet transform. *Geophysics*, *70*(6), P19–P25. https://doi.org/10.1190/1.2127113

Sun, S., Castagna, J. P., & Siegfried, R. W. (2002). Examples of wavelet transform time-frequency analysis in direct hydrocarbon detection. Paper presented in SEG Technical Program Expanded Abstracts 2002 (pp. 457–460). Society of Exploration Geophysicists. https://doi.org/10.1190/1.1817281

Tingdahl, K. M. (1999). *Improving seismic detectability using intrinsic directionality, Paper B194*. Earth Science Centre, Goteberg University.

Tingdahl, K. M., 2003. Improving seismic chimney detection using directional attributes. In M. Nikravesh, F. Aminzadeh, L.A. Zadeh (Eds.), *Soft computing and intelligent data analysis in oil exploration, developments in petroleum science* (Vol. *51*, pp. 157–173). Elsevier.

Tingdahl, K. M. & de Groot, P. F. (2003). Post-stack dip and azimuth processing. *Journal of Seismic Exploration*, *12*, 113–126.

Tingdahl, K. M. & de Rooij, M. (2005). Semi-automatic detection of faults in 3D seismic data. *Geophysical Prospecting*, *53*, 533–542. https://doi.org/10.1111/j.1365-2478.2005.00489.x

Tonn, R. (1991). The determination of the seismic quality factor Q from VSP data: A comparison of different computational methods: *Geophysical Prospecting*, *39*(1), 1–27. https://doi.org/10.1111/j.1365-2478.1991.tb00298.x

Tucker, P. M., & Yorston, H. J. (1973). *Pitfalls in seismic interpretation*. Society of Exploration Geophysicists.

Vernengo, L. & Trinchero, E. (2015). Application of amplitude volume technique attributes, their validations, and impact. *Leading Edge*, *34*, 1246–1253. https://doi.org/10.1190/tle34101246.1

3

BE AN INTERPRETER

The purpose of this chapter is to expose readers to a set of examples with a view to building and improving interpretation skills.

This chapter presents some exercises related to subsurface interpretation. The example cross-sections, each with a set of questions, provide an opportunity to practice recognizing geologic/structural features on seismic sections. Solutions for the interpretation tasks and answers to the numerical tasks are provided in Appendix C.

The answers for the numerical tasks are provided in Appendix B. Solutions to the interpretation tasks are also provided in Appendix B. This will be useful to the teachers and lab instructors.

Students, teachers, or instructors can download sample data sets from our institute server for hands-on practice. Note that the files are in SGY format. Software such as SeisSee, Seismic Trace Viewer (STV), Open Inventor SEG Y reader, GSEGYView, kogeo or SEGY Scout is needed to open these files and visualize the data. Separate seismic interpretation softwares such as PetrelTM, LandmarkTM, JasonTM, HRSTM, SeisEarthTM, DUG InsightTM, PaleoScanTM or OpendTectTM is then needed for interpreting the SEGY data files.

To access the files, go to http://14.139.225.217/ai_book_data/ai_book_data.zip. When prompted enter user1 as username and wihg248))! as password. The data will be downloaded as a zipped file. If you experience any difficulties accessing the files, please contact us at kumarchinmoy@gmail.com or chinmoy@wihg.res.in.

Meta-Attributes and Artificial Networking: A New Tool for Seismic Interpretation,
Special Publications 76, First Edition. Kalachand Sain and Priyadarshi Chinmoy Kumar.
© 2022 American Geophysical Union. Published 2022 by John Wiley & Sons, Inc.
DOI: 10.1002/9781119481874.ch03

3.1. Task 1

A seismic section is shown in Figure 3.1 from a 3D prospect of Taranaki Basin in New Zealand. The data is displayed using the SEG's American polarity convention, where an increase in acoustic impedance is represented by a peak (a positive amplitude black reflection) on seismic sections.

a) Interpret all the observed faults.
b) Which type of faults are these?
c) Do you see any shearing effect?
d) How many faults can be traced?
e) How many horst and graben structures can be interpreted?
f) Prepare a sketch of observed geological features.

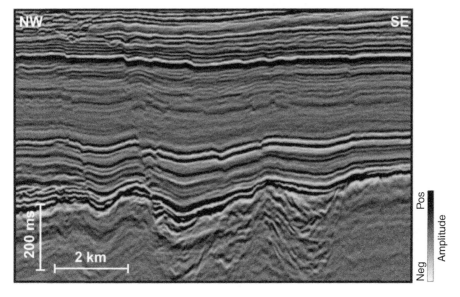

Figure 3.1 Example for Task 1.

3.2. Task 2

A seismic section is shown in Figure 3.2 from a 3D prospect of Taranaki Basin in New Zealand. The data is displayed using the SEG's American polarity convention, where an increase in acoustic impedance is represented by a peak (a positive amplitude black reflection) on seismic sections.

a)   Can you identify the Mass Transport Deposit (MTD)?
b)   Trace the basal shear surface (BSS) and top of MTD.
c)   Do you see any listric faults?
d)   If yes, map the faults.
e)   Prepare a sketch of observed geological features.

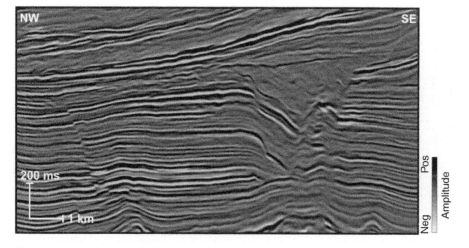

Figure 3.2   Example for Task 2.

3.3. Task 3

A seismic section is shown in Figure 3.3 from a 3D prospect of Taranaki Basin in New Zealand. The data is displayed using the SEG's American polarity convention, where an increase in acoustic impedance is represented by a peak (a positive amplitude black reflection) on seismic sections.

a)   Identify sigmoidal reflections.
b)   Do you observe onlap?
c)   Can you detect any parallel reflections?
d)   What are the trends of these reflections?
e)   Identify and trace unconformity.
f)   Do you see any channel signatures?
g)   Prepare a sketch of observed geological features.

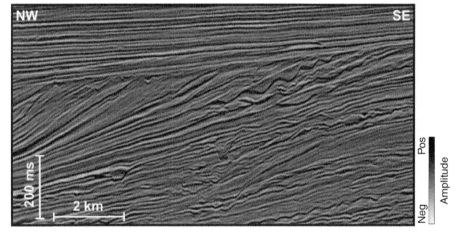

Figure 3.3   Example Task 3.

3.4. Task 4

A seismic section is shown in Figure 3.4 from a 3D prospect of Taranaki Basin in New Zealand. The data is displayed using the SEG's American polarity convention, where an increase in acoustic impedance is represented by a peak (a positive amplitude black reflection) on seismic sections.

  a)   Identify reflections associated with channel belts.
  b)   Trace the top and bottom of MTDs.
  c)   Do you observe any sediment peel-back within the MTD?
  d)   If yes, mark the effect.
  e)   Can you identify a ramp within the MTD zone?
  f)   Prepare a sketch of observed geological features.

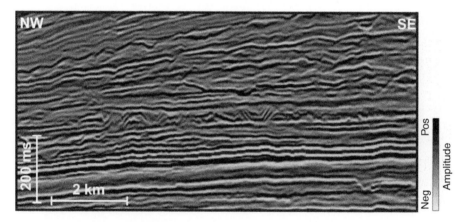

Figure 3.4   Example for Task 4.

3.5. Task 5

A seismic section is shown in Figure 3.5 from a 3D prospect of Taranaki Basin in New Zealand. The data is displayed using the SEG's American polarity convention, where an increase in acoustic impedance is represented by a peak (a positive amplitude black reflection) on seismic sections.

a)   What do you observe from the seismic cross-section?
b)   Do you see any magmatic sill(s)?
c)   If yes, mark all possible sills on the section.
d)   What type of structural geometry do they exhibit?
e)   Do you observe any fold(s)?
f)   If yes, what are the nature of fold(s)?
g)   Why do these structures develop?
h)   Prepare a sketch of observed geological features.

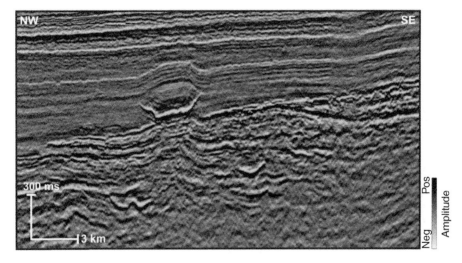

Figure 3.5   Example for Task 5.

3.6. Task 6

A seismic section is shown in Figure 3.6 from a 3D prospect of Taranaki Basin in New Zealand. The data is displayed using the SEG's American polarity convention, where an increase in acoustic impedance is represented by a peak (a positive amplitude black reflection) on seismic sections.

a) What do you observe from the seismic cross-section?
b) Do you see a DHI? Why does it occur?
c) Do you see any buried volcano?
d) If yes, can you mark the top of the volcano?
e) What relationship does the DHI have to the volcano?
f) What type of seismic attributes are preferred for interpreting buried volcanoes?

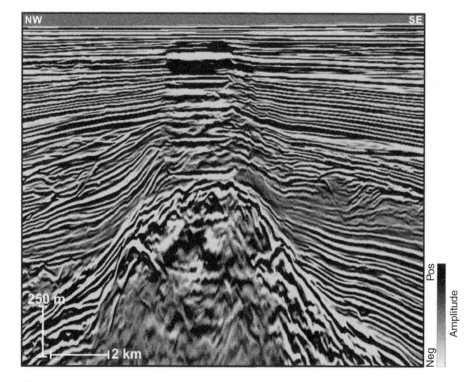

Figure 3.6 Example for Task 6.

3.7. Task 7

Assume a signal $x(t)$ such that $x(t) = \{0, 0, 1, 1, 1, 2, 3, 5\}$
a) Do you observe any noise bursts within the signal? What are those noise bursts?
b) Apply median filter to this signal.
c) What kind of optimum window length is preferred for removing noise bursts?

3.8. Task 8

Given an input signal sequence $x(t)$ such that $x(t) = \{0, 0, 1, 6\}$ and a kernel function $h(t)$ such that $h(t) = \{0, 0, 0, 1\}$.
a) What type of output signal is obtained when these two sequences are convolved?
b) Do you observe any noisy amplitudes within the output signal?
c) If yes, apply a 3-step median filter to the output signal.

3.9. Task 9

a) Define a Fourier series.
b) What type of transform should be used to convert a signal in frequency domain to the time domain?
c) Define a Hilbert Transform.
d) Define the mathematical equation to perform the convolution of two functions f and g.

3.10. Task 10

Given an input signal $x(t)$ and the corresponding impulse response $h(t)$.
$x(t) =$

4	5	6
1	2	3

$h(t) =$

1	1
1	1

a) Perform a 2D convolution.
b) What is the difference between convolution and correlation?
c) Why is correlation between the signals carried out?

Part II
Meta-Attributes

4

AN OVERVIEW OF META-ATTRIBUTES

The study of seismic attributes helps to elucidate the subsurface geologic body, but no single attribute will always correspond to a particular structure or feature. However, a hybrid-attribute can be designed by amalgamating a set of attributes into a single attribute which, in turn, can delimit a particular geologic feature with greater certainty. This chapter describes these hybrid attributes, termed meta-attributes, which have evolved recently to augment the interpretation of seismic data, especially for 3D seismic data.

4.1. Introduction

Seismic field data undergoes rigorous processing steps to produce a meaningful subsurface image in terms of geological structures or stratigraphic features (Yilmaz, 2001), which is interpreted for the exploration of hydrocarbons, identification of mineralized prospects, understanding geo-tectonic implications, etc. The image can be further enhanced and refined by employing state-of-the-art advanced processing tools with fewer uncertainties. Part I of this book provided an overview of the seismic attributes that have brought out revolutionary changes in the interpretation of subsurface features and properties from seismic data. Hereafter, we assume that readers have gained an appreciable idea of the historical evolution, types of seismic attributes, and their applications. All these are very important to remember, as these shall be used as inputs for generating what we call the hybrid- or meta-attributes.

Meta-Attributes and Artificial Networking: A New Tool for Seismic Interpretation,
Special Publications 76, First Edition. Kalachand Sain and Priyadarshi Chinmoy Kumar.
© 2022 American Geophysical Union. Published 2022 by John Wiley & Sons, Inc.
DOI: 10.1002/9781119481874.ch04

4.2. Meta-Attributes

Seismic attributes show the characteristics of geological features such as faults, fractures or folds; stratigraphic features such as pinch-outs, debris, slumps or channels; and intrusive elements such as sills, dykes, submarine volcanoes or reefs (Bahorich & Farmer, 1995; Barnes, 2003, 2016; Chopra & Marfurt, 2007, 2010; Reynolds et al., 2017; Tingdahl & de Groot, 2003). These attributes have been recently used as major inputs for delineating the 3D geometry of geologic features from seismic volume.

Suppose an interpreter finds faults or channels within the amplitude volume; they then like to quickly compute certain other attributes, e.g., coherency, curvature, dip magnitude, etc., to capture the signature of the said geologic features. Next the interpreter observes discontinuous structures (e.g., fault complexities) in the data that are associated with low coherency values. At the same time, they find that this attribute also indicates the meandering channel complexes. However, the interpreter might be interested in the interpretation of either the fault or the channel complex. This implies that a single attribute may hardly ever respond to a particular geologic object of interest (Kumar & Sain, 2018, 2020a; Kumar et al., 2019a, 2019b, 2019c; Sain & Kumar, 2021; Singh et al., 2016) and it will be difficult to discriminate geologic targets of different origins. Hence, the output may overlap with unwanted features or misinterpret subsurface. Thus, the interpretation becomes difficult unless a clear and distinct image of the geologic target is perceived. There is a need to come up with an optimal solution that may streamline the interpretation without any ambiguity.

Seismic attributes are computed using a set of algorithms from the data. In such a case, is it possible for an attribute to capture exactly a geologic target from the entire volume of data? To find an answer, a new attribute, defined as the hybrid- or meta-attribute, has been computed by combining a set of other attributes corresponding to a specific geologic target (Kumar & Sain, 2018, 2020a, 2020b; Kumar et al., 2019a, 2019b, 2019c; Singh et al., 2016). This concept has opened up a new dimension in seismic attributes for the advanced interpretation of 3D seismic data.

The first efforts in this direction were made by Aminzadeh and Chatterjee (1985), who combined several attributes to solve E & P problems by clustering analysis (in particular principal component analysis). Techniques such as regression analysis, principal component analysis, and clustering analysis have been used for multi-attribute analysis. However, such techniques could not incorporate the interpreter's familiarity with geology and tectonics, and thus acted as a black-box.

Meldahl et al. (1999, 2001) introduced a concept that formed the basis of designing a hybrid attribute. They suggested amalgamating several attributes and training them as input under the guidance of an interpreter's knowledge through artificial intelligence of neural networks to generate hybrid- or meta-attributes. This brought revolutionary changes to the interpretation domain of applied seismology. The most exciting part of the meta-attribute concept is the amalgamation of human intelligence with artificial intelligence to aid interpretation of a huge volume of data.

Hybrid attributes have been computed from 3D seismic data in several geologic environments that have prominently brought out the subsurface complexities, e.g., discontinuous structures, gas clouds and seepages, mass transport deposits, intrusive bodies like buried volcanoes, magmatic sill complexes, reefs, etc. (Brouwer et al., 2008; Connolly & Garcia, 2012; Ligtenberg, 2003; Kumar & Mandal, 2017; Kumar & Sain, 2018, 2020a, 2020b; Kumar et al., 2019a, 2019b, 2019c; Singh et al., 2016; Tingdahl & de Rooij, 2005). Meta-attributes are robust and much more efficient than individual attributes with regard to delimiting a subsurface geologic body. These hybrid attributes have also been used in discerning or characterizing reservoir properties such as porosity, saturation, and lithology (Aminzadeh et al., 2004) from seismic data.

4.3. Types of Meta-Attributes

Here we introduce a set of meta-attributes that have been computed to automate the process of interpretation of 3D seismic data: the Hydrocarbon Probability meta-attribute, the Chimney Cube meta-attribute, the Fault Cube meta-attribute, the Thinned Fault Cube meta-attribute, the Intrusion Cube meta-attribute, the Sill Cube meta-attribute, the Mass Transport Deposit Cube meta-attribute, and the Lithology meta-attribute.

4.3.1. Hydrocarbon Probability Meta-Attribute

The Hydrocarbon Probability (HP) meta-attribute (Aminzadeh & de Groot, 2005) is designed for the identification of plausible areas with hydrocarbon accumulations using the absorption-related seismic attribute, which is a measure of high-frequency loss due to transmission of signals through hydrocarbon reservoirs. The input data used for the computation of this meta-attribute could be angle-gathered data that shows variation of amplitudes with offsets. The training for generating the HP meta-attribute is based on the geologist's understanding of and familiarity with identifying the best possible locations within the data volume for the exploration of hydrocarbons. This interpretation could be further validated through several pieces of known borehole information in the prospective area.

4.3.2. Chimney Cube Meta-Attribute

Gas chimneys form chaotic disordered vertical disturbances on seismic data, causing weaker reflection amplitudes and discontinuous reflectors. Mapping chaotic reflections from seismic data and gleaning valuable information is a challenge in the hydrocarbon industry. Meldahl et al. (2001) and Connolly and Aminzadeh (2003) demonstrated that a gas chimney on the seismic data can be used as an efficient marker for hydrocarbons. This proxy can also be used for the detection of geo-hazards. Seismic characteristics associated with the chimney are low similarity, low amplitude, variable dip, and wipe-out zone due to high-frequency attenuation caused by scattering of seismic signals (Berndt et al., 2003; Brouwer et al., 2008; Connolly & Garcia, 2012; Ligtenberg, 2003; Petersen et al., 2010; Singh et al., 2016; Westbrook et al., 2008). Suitable seismic attributes, e.g., similarity, energy, dip variance, frequency, etc. (Brouwer et al., 2008; Connolly & Garcia, 2012; Singh et al., 2016), which respond to the chimney, can be amalgamated by a neural network and trained at example locations, randomly selected over a volume of data, to obtain a hybrid-attribute: the Chimney Cube (CC) meta-attribute. The interpreter's knowledge is crucial, as they decide all possible locations associated with the chimneys and non-chimneys within the data while training the system. The identification of gas chimneys provides a clue for the exploration of hydrocarbons and towards understanding the petroleum system of a region. The CC meta-attribute also acts as a guiding tool for identifying over-pressured zones, a prerequisite for avoiding drilling risks, if any (Heggland, 2004).

4.3.3. Fault Cube Meta-Attribute

Geologic faults are discontinuous structures, associated with the termination of reflectors, vertical disturbances, and discontinuities in seismic reflectors (Chopra & Marfurt, 2007; Tingdahl, 2003; Tingdahl & de Groot, 2003). These features are characterized by abrupt changes in dips of seismic reflectors and are generally oriented with different geometric shapes (e.g., curve, straight, and bend) (Chopra & Marfurt, 2007; Roberts, 2001; Tingdahl, 2003; Tingdahl & de Groot, 2003; Tingdahl & de Rooij, 2005). Moreover, fault zones are generally associated with the loss in signal energy and frequency content. Seismic attributes, e.g., similarity, coherency, curvature, dip angle variance, dip magnitude, azimuth, and energy attributes, which respond to faults, can be amalgamated by a neural system and trained by an interpreter's acquaintances to generate a hybrid attribute, defined as the Fault Cube (FC) meta-attribute. The interpreter plays an important role to decide possible example locations associated with the faults and non-fault locations within a small volume of data. Tingdahl and de Rooij (2005) presented a classic approach for generating fault meta-attribute from seismic data. Several other authors (Kluesner & Brothers, 2016; Kumar & Mandal, 2017; Mandal & Srivastava, 2018; Zheng et al., 2014) showed the extensive application of the FC meta-attribute for structural interpretation of seismic data. Readers should explore Chapters 8 and 9, which present a comprehensive study of generating FC meta-attributes through a combination of different sets of attributes and further demonstrate the process of generating the Thinned Fault Cube (TFC) meta-attribute, which has brought out sharp continuous fault images from seismic data.

4.3.4. Intrusion Cube Meta-Attribute

Buried volcanic systems are accompanied by igneous and sedimentary processes that may have both positive and negative impacts on the petroleum system: the maturity of source rocks; migration pathways for hydrocarbons; sealing and trapping mechanism; geothermal history of the basin, etc. (Allis et al., 1995; Farrimond et al., 1999; Planke et al., 2005; Rateau et al., 2013; Rohrman, 2007; Schutter, 2003; Stagpoole & Funnell, 2001; Sun et al., 2014; Zou, 2013). These subsurface complexities sometimes change the porosity and permeability of the reservoirs and pose risks for exploration and production. Such complex geological systems should be examined critically to avoid any hazards. High-quality seismic data is a reliable tool for delineating the detailed architecture of these buried geological structures (Cartwright & Hansen, 2006; Polteau et al., 2008; Rateau et al., 2013; Smallwood & Maresh, 2002; Sun et al., 2014; Thomson & Schofield, 2008).

In seismic reflection data, intrusive features such as the buried volcanoes, sills, dykes, etc. are usually well-imaged due to their high-amplitude character (which is a function of density and velocity), abrupt lateral terminations, and complex geometries within the host-rock strata (Alves et al., 2015; Planke et al., 2005; Smallwood & Maresh, 2002). Volcanoes have a tendency to pierce vertically through the host sedimentary succession. Internally, they are associated with distorted and chaotic reflections and dissimilar seismic events. Over the crater and along the flanks, they are associated with high amplitudes due to larger impedance contrast with the surrounding sedimentary rocks (Bischoff et al., 2017; Infante-Paez & Marfurt, 2017). While piercing through the host geologic formations, these features tend to possess variable dips. Seismic energy, while passing through these features, gets diminished. The flanks and the crater of the volcano are linked with high coherent energy. Sills and dykes generally act as conducive pathways for the movement of magma into the overlying sedimentary units. Sills, concordant intrusive bodies, commonly sub-horizontal with a gentle inclination and cross-cutting stratigraphy, exhibit a major impact on basin history and petroleum system (Cartwright & Hansen, 2006; Smallwood & Maresh, 2002; Thomson & Hutton, 2004). On the seismic section, these appear as saucer-shaped and gently inclined high amplitude features, which are discontinuous in nature. However, dykes vertically intrude into the overlying younger strata and exhibit steep dips. Thus, the reflection strength, dip angle variance, energy gradient, azimuth, and similarity attributes, which respond to the intrusives, when fused together by a neural training system, generate a hybrid-attribute called the Intrusion Cube (IC) meta-attribute. This attribute correctly captures the extension and distribution of such complex plumbing system from seismic data. Kumar et al. (2019b) made a first documentation of IC meta-attribute to explain the plumbing system (consisting of the buried volcano, sill networks, dyke swarms, magmatic ascent) from seismic data. Readers are directed to Chapter 13, which demonstrates the application of the IC meta-attribute to the interpretation of magma plumbing system from seismic data.

4.3.5. Sill Cube Meta-Attribute

Understanding the evolution, emplacement processes, and geometries of magmatic sill complexes is essential for evaluating magmatic fluid plumbing system and their impact on petroleum system (Cartwright & Hansen, 2006; Galland, 2012; Millet et al., 2016; Muirhead et al., 2016; Rateau et al., 2013; Schofield et al., 2017; Senger et al., 2017). Conventionally, magmatic sills are interpreted as tabular intrusive rocks with concordant surfaces showing concave upwards cross-sectional geometries and discordant limbs (Allaby & Allaby, 1999; Hansen et al., 2004). These concordant intrusive bodies, commonly sub-horizontal with gentle inclination, cross-cut the stratigraphy, and exhibit a major impact on basin history and the petroleum system (Cartwright & Hansen, 2006; Omosanya, 2018; Smallwood & Maresh, 2002; Thomson & Hutton, 2004). Such geologic features help in transporting magmatic fluids into the overlying stratigraphic units. The magmatic sills can be detected by high amplitude and entropy character (a measure of disorder or complexity) due to larger impedance contrast with the surrounding sedimentary rocks. Moreover, they possess very good contrast with the host sedimentary successions (Chopra & Marfurt, 2007). These seismic characteristics provide suitable attributes, e.g., reflection strength, texture contrast, texture entropy, etc., which can be combined into one attribute through neural training based on an interpreter's guidance to generate a hybrid attribute called the Sill Cube (SC) (or sill) meta-attribute. Kumar et al. (2019a) presented a classic example of SC meta-attribute for the interpretation of magmatic sill complexes from seismic data, acquired in indifferent sedimentary basins. Moreover, this work also demonstrates the merit of meta-attribute computation compared to other interpretation techniques, e.g., geo-body extraction and red-green-blue (RGB) color blending. Readers are requested to explore Chapter 12 to learn more about the concept of SC meta-attributes and their applications for the interpretation of magmatic sills from seismic reflection data.

4.3.6. Mass Transport Deposit Cube Meta-Attribute

Mass transport deposits (MTDs) form an essential component of the slope system and occur in different tectonic and depositional settings. MTDs vary considerably in shape, size, and internal structure (Panpichityota et al., 2018). The advent of 3D seismic technology has led to detailed analysis and visualization of MTDs from seismic data (Martinez et al., 2005). Seismic attributes such as the dip angle variance, dip magnitude, azimuth, similarity, and energy, sensitive to MTDs, can be amalgamated by a neural network based on an interpreter's judgement on MTD-yes and MTD-no objects to generate a meta-attribute called the MTD Cube (MTDC) meta-attribute. Kumar and Sain (2020b) demonstrated the computation of the MTDC meta-attribute and illustrated its efficacy of interpretation over other conventional approaches. Readers should explore Chapter 14 to learn more about the computation and usage of the MTDC meta-attribute for the interpretation of MTDs from seismic data.

4.3.7. Lithology Meta-Attribute

In all of the above-described meta-attributes, the concept was restricted to two objectives i.e., target-yes and target-no. However, the Lithology meta-attribute, as described by Aminzadeh et al. (2004), is a three-class problem associated with three different objectives. The Lithology meta-attribute is a hybrid attribute that captures different Lithology (e.g., channel, levee, silt-shale, etc.) as desired by an interpreter. To understand more about Lithology meta-attributes, developed through the neural network tool for the characterization of subsurface reservoirs, readers may explore the works of de Groot (1995, 1999); Schuelke et al. (1997); Aminzadeh et al. (2004); Wong et al. (2002); and Sandham et al. (2003).

4.4. Summary

Meta-attributes have been very useful for advanced interpretation of 3D seismic data in terms of geologic features and reservoir properties. Apart from having the ability to blend several seismic attributes, the concept of the meta-attribute has allowed the scientific community to train the system by incorporating an interpreter's familiarity with a small segment of data, which can generate a reliable outcome with much-reduced uncertainties. Having explored the concept of meta-attributes, the next chapter turns to look at artificial neural networks.

References

Allaby, A., & Allaby, M. (1996). Dictionary of Earth Sciences. Oxford University Press.
Allis, R. G., Armstrong, P. A. & Funnell, R. H. (1995). Implications of high heat flow anomaly around New Plymouth, North Island, New Zealand. New Zealand Journal of Geology and Geophysics, 38, 121–130.
Allis, R. G., Armstrong, P. A., & Funnell, R. H. (1995). Implications of a high heat flow anomaly around New Plymouth, North Island, New Zealand. New Zealand Journal of Geology and Geophysics, 38(2), 121–130. https://doi.org/10.1080/00288306.1995.9514644

Alves, T.M., Omosanya, K.D. & Gowling, P. (2015). Volume rendering of enigmatic high-amplitude anomalies in southeast Brazil: A workflow to distinguish lithologic features from fluid accumulations. *Interpretation, 3,* A1–A14. https://doi.org/10.1190/INT-2014-0106.1

Aminzadeh, F., and Chatterjee, S. (1985). Applications of clustering in exploration seismology. *Geoexploration, 23,* 147–159. https://doi.org/10.1016/0016-7142(84)90028-0

Aminzadeh, F., de Groot, P., Wilkinson, D. (2004) Soft Computing for qualitative and quantitative seismic object and reservoir property prediction, Part 1, Neural network applications, Part 2, Fuzzy logic applications, Part 3, Evolutionary computing and other aspects of soft computing. *First Break, 22,* 49–54, 69–78, 107–116. https://doi.org/10.3997/1365-2397.22.6.25903

Aminzadeh, F., & de Groot, P. (2005). A neural networks based seismic object detection technique. Paper presented in SEG Technical Program Expanded Abstracts (pp. 775–778). Society of Exploration Geophysicists. https://doi.org/10.1190/1.2144442

Aminzadeh, F., & De Groot, P. (2006). *Neural networks and other soft computing techniques with applications in the oil industry.* EAGE Publications.

Bahorich, M., & Farmer, S. (1995). 3-D seismic discontinuity for faults and stratigraphic features: The coherence cube. *Leading Edge, 14,* 1053–1058. https://doi.org/10.1190/1.1437077

Barnes, A. E., 2003. Shaded relief seismic attribute. *Geophysics, 68,* 1281–1285. https://doi.org/10.1190/1.1598120

Barnes, A. E. (2016). *Handbook of poststack seismic attributes.* Society of Exploration Geophysicists. https://doi.org/10.1190/1.9781560803324

Bischoff, A. P., Andrew, N. & Beggs, M. (2017). Stratigraphy of architectural elements in a buried volcanic system and implications for hydrocarbon exploration. *Interpretation, 5,* SK141–159. https://doi.org/10.1190/INT-2016-0201.1

Berndt, C., Bunz, S. & Mienert, J. (2003). Polygonal fault systems on the Mid-Norwegian margin: a long-term source for fluid flow. In Rensbergen, P.V., Hills, R., Maltman, A., Morley, C. (Eds.), *Origin, processes, and effects of subsurface sediment mobilization on reservoir to regional scale* (pp. 283–290). Geological Society of London, Special Publication. https://doi.org/10.1144/GSL.SP.2003.216.01.18

Brouwer, F.G.C., Welsh, A., Connolly, D.L., Selva, C., Curia, D. & Huck, A. (2008). High Frequencies Attenuation and Low Frequency Shadows in Seismic Data Caused by Gas Chimneys, Onshore Ecuador. Paper presented in 70th EAGE Conference and Exhibition incorporating SPE EUROPEC 2008. https://doi.org/10.3997/2214-4609.20147600

Cartwright, J., & Møller Hansen, D. (2006). Magma transport through the crust via interconnected sill complexes. *Geology, 34*(11), 929–932. https://doi.org/10.1130/G22758A.1

Chopra, S., & Marfurt, K. J. (2007). *Seismic attributes for prospect identification and reservoir characterization.* SEG, Tulsa.

Chopra, S., & Marfurt, K. J. (2010). Integration of coherence and volumetric curvature images. *The Leading Edge, 29*(9), 1092–1107. https://doi.org/10.1080/00288306.1995

Connolly, D., & Aminzadeh, F. (2003). Exploring for Deep Gas in the Gulf of Mexico Shelf and Deepwater Using Gas Chimney Processing. *Gulf Coast Association of Geological Societies Transactions, 53,* 135–142

Connolly, D. & Garcia, R. (2012). GEOLOGY & GEOPHYSICS-Tracking hydrocarbon seepage in Argentina's Neuquén basin. World Oil, 101–104.

de Groot, P.F.M. (1995). Seismic reservoir characterisation employing factual and simulating wells [Doctoral thesis, TU Delft]. Delft University Press. http://resolver.tudelft.nl/uuid:a596871e-daf9-4ac6-a519-dec802151162

de Groot, P.F.M. (1999). *Seismic reservoir characterisation using artificial neural networks.* Paper presented at 19th Mintrop Seminar, 16–18.

Farrimond, P., Bevan, J. C., & Bishop, A. N. (1999). Tricyclic terpane maturity parameters: response to heating by an igneous intrusion. *Organic Geochemistry, 30*(8), 1011–1019. https://doi.org/10.1016/S0146-6380(99)00091-1

Galland, O. (2012). Experimental modelling of ground deformation associated with shallow magma intrusions. *Earth and Planetary Science Letters, 317,* 145–156. https://dx.doi.org/10.1016/j.epsl.2011.10.017

Hansen, D. M., Cartwright, J. A., & Thomas, D. (2004). 3D seismic analysis of the geometry of igneous sills and sill junction relationships. *Geological Society, London, Memoirs, 29*(1), 199–208. https://doi.org/10.1144/GSL.MEM.2004.029.01.19

Heggland, R. (2004). Definition of geohazards in exploration 3-D seismic data using attributes and neural-network analysis. *AAPG Bulletin, 88,* 857–868. https://doi.org/10.1306/02040404019

Infante-Paez, L., & Marfurt, K. J. (2017). Seismic expression and geomorphology of igneous bodies: A Taranaki Basin, New Zealand, case study. *Interpretation, 5*(3), SK121-SK140. https://doi.org/10.1190/INT-2016-0244.1

Kluesner, J. W., & Brothers, D. S. (2016). Seismic attribute detection of faults and fluid pathways within an active strike-slip shear zone: New insights from high-resolution 3D P-Cable™ seismic data along the Hosgri Fault, offshore California. *Interpretation, 4*(1), SB131–SB148. https://doi.org/10.1190/INT-2015-0143.1

Kumar P.C., & Mandal, A. (2017). Enhancement of fault interpretation using multi-attribute analysis and artificial neural network (ANN) approach: A case study from Taranaki Basin, *New Zealand. Exploration Geophysics, 49*(3), 409–424. https://doi.org/10.1071/EG16072

Kumar, P. C., & Sain, K. (2018). Attribute amalgamation-aiding interpretation of faults from seismic data: An example from Waitara 3D prospect in Taranaki basin off New Zealand. *Journal of Applied Geophysics, 159,* 52–68. https://doi.org/10.1016/j.jappgeo.2018.07.023

Kumar, P.C., Omosanya, K. O., Sain, K. (2019a). Sill Cube: An automated approach for the interpretation of magmatic sill complexes on seismic reflection data. *Marine and Petroleum Geology, 100,* 60–84. https://doi.org/10.1016/j.marpetgeo.2018.10.054

Kumar, P.C., Sain, K. & Mandal, A. (2019b). Delineation of a buried volcanic system in Kora prospect off New Zealand using artificial neural networks and its implications. *Journal of Applied Geophysics, 161,* 56–75. https://doi.org/10.1016/j.jappgeo.2018.12.008

Kumar, P.C., Omosanya, K.O., Alves, T. & Sain, K. (2019c). A neural network approach for elucidating fluid leakage along hard-linked normal faults. *Journal of Marine and Petroleum Geology, 110,* 518–538. https://doi.org/10.1016/j.marpetgeo.2019.07.042

Kumar, P.C. & Sain, K. (2020a). Interpretation of magma transport through saucer sills in shallow sedimentary strata using an automated machine learning approach. *Tectonophysics, 789,* 228541, 1–16. https://doi.org/10.1016/j.tecto.2020.228541

Kumar, P.C. & Sain, K. (2020b). A machine learning tool for interpretation of Mass Transport Deposits from seismic data. *Scientific Reports, 10*(4134), 1–10. https://doi.org/10.1038/s41598-020-71088-6

Ligtenberg, J. H. (2003) Unravelling the petroleum system by enhancing fluid migration paths in seismic data using a neural network based pattern recognition technique. *Geofluids, 4*, 255–261. https://doi.org/10.1046/j.1468-8123.2003.00072.x

Mandal, A., & Srivastava, E. (2018). Enhanced structural interpretation from 3D seismic data using hybrid attributes: New insights into fault visualization and displacement in Cretaceous formations of the Scotian Basin, offshore Nova Scotia. *Marine and Petroleum Geology, 89*, 464–478. https://doi.org/10.1016/j.marpetgeo.2017.10.013

Martinez, J. F., Cartwright, J., & Hall, B. (2005). 3D seismic interpretation of slump complexes: examples from the continental margin of Israel. *Basin Research, 17*(1), 83–108. https://doi.org/10.1111/j.1365-2117.2005.00255.x

Meldahl, P., Heggland, R., Bril, B., & de Groot, P. (1999). The chimney cube, an example of semi-automated detection of seismic objects by directive attributes and neural networks: Part I; methodology. In *SEG Technical Program Expanded Abstracts* (pp. 931–934). SEG. https://doi.org/10.1190/1.1821263

Meldahl, P., Heggland, R., Bril, B., & de Groot, P. (2001). Identifying faults and gas chimneys using multiattributes and neural networks. *The Leading Edge, 20*(5), 474–482. https://doi.org/10.1190/1.1438976

Millett, J.M., Wilkins, A.D., Campbell, E., Hole, M.J., Taylor, R.A., Healy, D., Jerram, D. A., Jolley, D.W., Planke, S., Archer, S.G., Blischke, A., 2016. The geology of offshore drilling through basalt sequences: Understanding operational complications to improve efficiency. *Marine and Petroleum Geology, 77*, 1177–1192. https://doi.org/10.1016/j.marpetgeo.2016.08.010

Muirhead, J. D., Van Eaton, A. R., Re, G., White, J. D., Ort, M. H., 2016. Monogenetic volcanoes fed by interconnected dikes and sills in the Hopi Buttes volcanic field, Navajo Nation, USA. *Bulletin of Volcanology, 78*, 1–16. https://doi.org/10.1007/s00445-016-1005-8

Omosanya, K. O. (2018). Episodic fluid flow as a trigger for Miocene-Pliocene slope instability on the Utgard High, Norwegian Sea. *Basin Research, 30*(5), 942–964. https://doi.org/10.1111/bre.12288

Panpichityota, N., Morley, C. K., & Ghosh, J. (2018). Link between growth faulting and initiation of a mass transport deposit in the northern Taranaki Basin, New Zealand. *Basin Research, 30*(2), 237–248. https://doi.org/10.1111/bre.12251

Petersen, C.J., Bunz, S., Hustoft, S., Mienert, J. & Klaeschen, D. (2010). High-resolution P-Cable 3D seismic imaging of gas chimney structures in gas hydrate sediments of an Arctic sediment drift. *Marine and Petroleum Geology, 27*, 1981–1994. https://doi.org/10.1016/j.marpetgeo.2010.06.006

Planke, S., Rasmussen, T., Rey, S. S. & Myklebust, R. (2005). Seismic characteristics and distribution of volcanic intrusions and hydrothermal vent complexes in the Vøring and Møre basins. In Petroleum Geology Conference series, 6 (p. 833–844). Geological Society, London. https://doi.org/10.1144/0060833

Polteau, S., Mazzini, A., Galland, O., Planke, S., & Malthe-Sørenssen, A. (2008). Saucer-shaped intrusions: Occurrences, emplacement and implications. *Earth and Planetary Science Letters, 266*(1–2), 195–204. https://doi.org/10.1016/j.epsl.2007.11.015

Rateau, R., Schofield, N. & Smith, M. (2013). The potential role of igneous intrusions on hydrocarbon migration, West Shetland. *Petroleum Geoscience, 19*, 259–272. https://doi.org/10.1144/petgeo2012-035

Reynolds, P., Holford, S., Schofield, N., & Ross, A. (2017). The shallow depth emplacement of mafic intrusions on a magma-poor rifted margin: An example from the Bight

Basin, southern Australia. *Marine and Petroleum Geology*, *88*, 605–616. https://doi.org/10.1016/j.marpetgeo.2017.09.008

Roberts, A. (2001). Curvature attributes and their application to 3-D interpreted horizons. *First Break*, *19*, 85–100. https://doi.org/10.1046/j.0263-5046.2001.00142.x

Rohrman, M. (2007). Prospectivity of volcanic basins: Trap delineation and acreage de-risking. *AAPG bulletin*, *91*(6), 915–939. https://doi.org/10.1306/12150606017

Sain, K. & Kumar, P.C. (2021). Seismic, neural intelligence to artificial intelligence for seismic interpretation. In H.K. Gupta (Ed.), *Encyclopedia of solid earth geophysics* (2nd Edition). Springer Publications, in press.

Sandham, W., Leggett, L., Aminzadeh. F. (2003). *Applications of artificial neural networks and fuzzy logic*. Kluwer Academic Publisher. Introduction.

Schofield, N., Holford, S., Millett, J., Brown, D., Jolley, D., Passey, S.R., Muirhead, D., Grove, C., Magee, C., Murray, J., Hole, M., 2017. Regional magma plumbing and emplacement mechanisms of the Faroe-Shetland Sill Complex: implications for magma transport and petroleum systems within sedimentary basins. *Basin Res. 29*, 41–63. https://doi.org/10.1111/bre.12164

Schuelke, J.S., Quirein, J.A., Sarg, J.F., Altany, D.A. & Hunt, P.E. (1997). Reservoir architecture and porosity distribution, Pegasus field, West Texas—An integrated sequence stratigraphic– seismic attribute study using neural networks, Paper presented in SEG Annual Meeting, Dallas, Texas. https://doi.org/10.1190/1.1886093

Schutter, S. R., (2003). Hydrocarbon occurrence and exploration in and around igneous rocks: *Geological Society of London, Special Publications*, *7–33*. https://doi.org/10.1144/GSL.SP.2003.214.01.02

Senger, K., Millett, J., Planke, S., Ogata, K., Eide, C.H., Festøy, M., Galland, O., Jerram, D.A., 2017. Effects of igneous intrusions on the petroleum system: a review. *First Break*, *35*, 47–56. https://doi.org/10.3997/1365-2397.2017011

Singh, D., Kumar, P.C. & Sain, K. (2016). Interpretation of gas chimney from seismic data using artificial neural network: A study from Maari 3D prospect in the Taranaki basin, New Zealand. *Journal of Natural Gas Science and Engineering*, *36*, 339–357. https://doi.org/10.1016/j.jngse.2016.10.039

Smallwood, J.R. & Maresh, J. (2002). The properties, morphology and distribution of igneous sills: modelling, borehole data and 3D seismic from the Faroe-Shetland area. *Geological Society London, Special Publication*, *197*, 271–306. https://doi.org/10.1144/GSL.SP.2002.197

Stagpoole, V., & Funnell, R. (2001). Arc magmatism and hydrocarbon generation in the northern Taranaki Basin, New Zealand, *Petroleum Geoscience*, *7*, 255–267. https://doi.org/10.1144/petgeo.7.3.255

Sun, Q., Wu, S., Cartwright, J., Wang, S., Lu, Y., Chen, D. & Dong, D. (2014), Neogene igneous intrusions in the northern South China Sea: Evidence from high-resolution three-dimensional seismic data. *Marine and Petroleum Geology*, *54*, 83–95. https://doi.org/10.1016/j.marpetgeo.2014.02.014

Thomson, K. & Hutton, D. (2004). Geometry and growth of sill complexes: insights using 3D seismic from the North Rockall Trough. *Bulletin of Volcanology, 66*, 364–375. https://doi.org/10.1007/s00445-003-0320-z

Thomson, K. & Schofield, N. (2008). Lithological and structural controls on the emplacement and morphology of sills in sedimentary basins. *Geological Society London Special Publication*, *302*, 31–44. https://doi.org/10.1144/SP302.3

Tingdahl, K. M., (2003). Improving seismic chimney detection using directional attributes. In M. Nikravesh, F. Aminzadeh, L.A. Zadeh (Eds.), *Soft computing and intelligent data analysis in oil exploration. Developments in petroleum science*, Vol. *51*, pp. 157–173). Elsevier. https://doi.org/10.1016/S0920-4105(01)00090-0

Tingdahl, K. M., & de Groot, P. F. (2003). Post-stack dip and azimuth processing. *Journal of Seismic Exploration, 12*(2), 113–126.

Tingdahl, K. M. & de Rooij, M. (2005). Semi-automatic detection of faults in 3D seismic data; *Geophysical Prospecting, 53*, 533–542. https://doi.org/10.1111/j.1365-2478.2005.00489.x

Westbrook, G.K., Exley, R., Minshull, T., Nouze, H., Gailler, A., Jose, T., Ker, S. & Plaza, A., (2008). High-resolution 3D Seismic Investigations of Hydrate-bearing Fluid-escape Chimneys in the Nyegga Region of the Vøring Plateau, Norway. In *Proceedings of the 6th International Conference on Gas Hydrates*, Vancouver, British Colombia, July 6–10, pp. 8.

Wong, P.M., Aminzadeh, F., Nikravesh, M. (Eds.) (2002). *Soft Computing for Reservoir Characterisation and Modeling, Studies in Fuzziness and Soft Computing*. Physica-Verlag, Springer-Verlag. https://doi.org/10.1007/978-3-7908-1807-9

Yilmaz, Öz (2001). *Seismic data analysis: Processing, inversion, and interpretation of seismic data*. S. M. Doherty (Ed.), Society of Exploration Geophysicists, Tulsa, OK, USA. https://doi.org/10.1190/1.9781560801580

Zheng, Z.H., Kavousi, P., & Di, H.B. (2014). Multi-attributes and neural network-based fault detection in 3D seismic interpretation. *In Advanced Materials Research, Trans Tech Publications*, vol. *838*, pp. 1497–1502. https://doi.org/10.4028/www.scientific.net/AMR.838-841.1497

Zou, C. (2013). *Volcanic reservoirs in petroleum exploration*. 1st ed. Elsevier. https://doi.org/10.1016/C2011-0-06248-8

5

AN OVERVIEW OF ARTIFICIAL NEURAL NETWORKS

A human analyst struggles to take a quick decision without any mistakes and solve a complex problem with a large volume of data. Machine intelligence, under the guidance of human intelligence, can be used to ease the process. This chapter provides an overview of Artificial Neural Network (ANN) technology and elaborates the mathematical structure of an ANN. It explains how a machine can be trained, the different types of ANN, and how this approach can be extended to the interpretation of seismic data.

5.1. Introduction

The human brain is an expert system for analyzing patterns and decision-making. For example, ask someone to identify their family members in a crowd and they will do it quickly. However, ask them to perform specific quantitative tasks, such as identify the weight and height of each person in the crowd, or calculate how much property, bank balance, land area, or vehicles each of them possesses, the computations become difficult. However, if a machine (e.g., calculator or a computer) is programmed to do such tasks, it provides the quantitative solutions rapidly and easily. A machine has the ability to work continuously and answer questions with great accuracy. When such machines are trained to work under the guidance of human intelligence, it is possible to achieve optimum solutions to large and complex problems. This book demonstrates the possibilities of

Meta-Attributes and Artificial Networking: A New Tool for Seismic Interpretation,
Special Publications 76, First Edition. Kalachand Sain and Priyadarshi Chinmoy Kumar.
© 2022 American Geophysical Union. Published 2022 by John Wiley & Sons, Inc.
DOI: 10.1002/9781119481874.ch05

integrating human intelligence with machine intelligence for the interpretation of a large volume of 3D seismic data in a complex environment.

The human brain, consisting of a dense set of neural networks (NNs), continuously receives input signals from different sensory organs and processes them to derive meaningful information. Similarly, a computer can be programmed through a set of mathematical equations and algorithms to solve complex problems. This marks the evolution of computational or artificial neural networks that resemble a biological neural system. Thus, Artificial NNs (ANNs) are information processing setups motivated by the working environment of a biological neural system. Such a system comprises a large number of interconnected processing elements that operate mutually to solve a particular problem of interest.

During the past decade, NNs have matured and found numerous applications in different domains including military, aerospace, automotive, electronics, telecommunications, speech recognition, finance, and medicine. Most recently, artificial intelligence technology has been very useful in containing and combatting the Covid-19 pandemic: screening, tracking, and predicting current and future patients; assisting in early detection and diagnosis of infection; and supporting the development of drugs and vaccines and the reduction of workloads of healthcare workers (Vaishya et al., 2020).

The application of this technique in geophysics has grown over the last few decades. NNs have been used in a variety of geoscience problems, such as waveform recognition and first beak picking (McCormack et al., 1993; Murat & Rudman, 1992), electromagnetics (Poulton et al., 1992), magnetotellurics (Zhang & Paulson, 1997), seismic inversion problems (Calderón-Macías, 1999; Langer et al., 1996; Röth & Tarantola, 1994), petrophysical analysis (Huang et al., 1996), trace editing (McCormack et al., 1993), seismic deconvolution (Calderón-Macías et al., 1997; Wang & Mendal, 1991, 1992), and reservoir characterization (Aminzadeh et al., 2004; Nikravesh et al., 2003; Sandham et al., 2003; Tonn, 2002; Wong et al., 2002). More recently, the application of NNs has been extended to the delimitation of subsurface geologic features from 3D seismic data (Kumar & Mandal, 2017; Kumar & Sain, 2018, 2020; Kumar et al., 2019a, 2019b, 2019c; Sain & Kumar, 2021; Singh et al., 2016). Readers may also have a look at the review articles (Chentouf et al., 1997; Hush & Horne, 1993; McCormack, 1991; Poulton, 2001; Van der Baan & Jutten, 2000).

In summary, ANNs are considered to be the most effective means of interpretation for the following reasons:

- *Faster*: ANNs are strong enough to handle large datasets, process them rapidly, and extract appropriate information with limited human intervention.
- *Robust and Fault-Tolerant:* ANNs can lever several unexpected situations and extrapolate them into new circumstances by amalgamating information from several other domains.

- *Flexibility and Adaptive Learning:* ANNs can adjust to different new environments and thus can learn and produce a reliable outcomes.
- *Automatization:* After being fully trained by human intelligence based on a small segment of data volume, ANNs can predict the target automatically from the entire set of data volume.

5.2. Historical Evolution

In 1943 the paper by McCulloch and Pitts entitled "The logical calculus of ideas immanent in nervous activity," showcased for the first time that a set of mathematical functions can mimic the architecture of a human neural system. The neural model is based on the following assumptions:

- A neuron is a binary device, and its value can be either 0 or 1.
- Each neuron is associated with a threshold limit that the summation of inputs must surpass before the neuron starts to compute an output.
- Both excitatory and inhibitory connections weights ($w = \pm 1$) act as pathways for the neuron to receive inputs.
- There is a time interval for the neurons to respond to the activity of weights.
- If no inhibitory consequences are active, the neuron can monitor if the sum meets or exceeds its threshold.

The neural model of McCulloch and Pitts (1943) could compute the finite logical expression and is a simple threshold logical unit. Such a model infers that the human brain is a powerful logical system as their model is inspired by the concepts of neurophysiology.

Hebb (1949) made a breakthrough in the domain of neural intelligence through his book entitled *Organization of Behavior* and introduced what is known as "Hebbian Learning." According to this, if the axon of an input neuron is near enough to excite a target neuron or if it precisely takes part in firing the target neuron, some growth processes take place in one or both cells to increase the efficiency of the input neuron's stimulation. However, Hebb (1949) was not able to express his ideas mathematically; instead the neural community today uses his ideology of weighted connections that maintain a relationship among the processing elements within the network. Furthermore, Rochester et al. (1956) were able to successfully test Hebb's theory in computer simulations, which revolutionized neural network research. The period between 1946 and 1957s the evolution of neuro-computational techniques. Rosenblatt (1958) introduced the theory of the Perceptron into the neural system, which comprises three layers: an input layer consisting of retinal units, a middle layer consisting of association units, and an output layer consisting of response units. Each layer is connected through a set of randomized connections to what we call the synaptic weights. These connections could be modified in due course through a reinforcement mechanism. Rosenblatt's (1958) theory of the Perceptron suffered from limitations, and had the ability to solve classification problems pertaining to linearly separable classes only. However, issues in higher dimensional spaces are associated with non-linearity and thus Perceptron theory cannot always find a reliable solution. Rosenblatt added that Perceptron theory is very similar to a brain-damaged patient, where it can recognize qualitative features (color, shape, size) but struggles to find a relationship between these features (e.g., the name of an object exhibiting a bowl shape). Rosenblatt's (1958) neuro machine was called ADAptive LInear NEtwork (ADALINE), and showed good command over generalization ability and poor performance over abstraction capability.

Widrow and Hoff (1960) introduced the MADALINE (Many ADALINEs) neuro machine, which introduced the Delta learning rule that can change and update connection weights to minimize the error between the desired and calculated responses, again for linearly separable problems. Later, Caianiello (1961) modified the McCulloch and Pitts neural model and incorporated the neural processing of time-varying data sequences. Several ups and downs in the field of neural network research were observed during the 1960s to 1980s. A plethora of algorithms and new theories were developed from the 1990s onwards that redefined and restructured neural networks, which are being utilized in many science and engineering fields. Applications of neural networks demonstrate significant success over industrial challenges such as productivity, cost reductions, management & monitoring and improved qualitative output.

5.3. Biological Neuron vs Mathematical Neuron

5.3.1. Biological Neuron

The preceding sections provided an overview of neural networks along with their historical evolution. It is particularly interesting to understand the biological neuron (BN) from which the concept of the mathematical neuron (MN) evolved. The human brain is made up of billions of interconnected neurons through which signals are transmitted. The neurons generate electrical signals that are frequency coded instead of amplitude coded (Poulton, 2001). The central part of the human neural system (HNS), called the soma, contains a solution rich in potassium ions rather than sodium ions. However, the fluid surrounding the cell body is richer in sodium ions than potassium ions. Thus, a potential difference (PD) of 70 mV across the cell membrane is created by the concentration gradient (Fischbach, 1992; Poulton, 2001), the change of which alters the PD and activates the neuron.

Structurally, a biological neuron (Figure 5.1) mainly consists of three parts: the dendrites, the soma, and the axon. A neuron receives input signals from other neurons connected to its dendrites by synapses. As the distance from the synapses

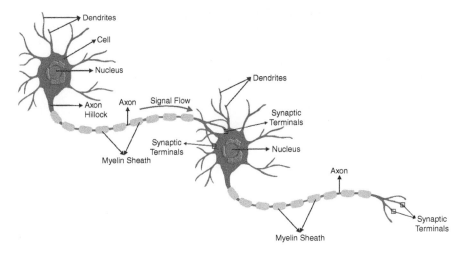

Figure 5.1   A typical architecture of the biological neuron within the human nervous system.

to the soma (the cell body) increases, the signal gets attenuated. Moreover, synapses control the connection strength between the soma and the dendrites. The weaker the synaptic strength, the poorer the signal transmission from the dendrite to the soma. Once the soma collects the input signal, it integrates them and then activates an output depending on total input. The output signal is then transmitted by the axon to other neurons by the synapses (Van der Baan & Jutten, 2000). It is important to know what controls the propagation of signal through the axon to other neurons. It is the activation that functions when the PD between the internal part of the cell and the external environment exceeds the threshold limit. The incoming signals are generally summed and accumulated within the soma for a short duration, and the output is fired to the axon. This process of adding incoming signals and making a check for the threshold is the primary motivation for the McCulloch and Pitts (1943) mathematical neuron, and the inclusion of a time constant during the summing and accumulation process is the motivation for the Caianiello neuron.

5.3.2. Mathematical Neuron

A mathematical neuron, also known as the perceptron (Rosenblatt, 1958), possesses a structure similar to that of a biological neuron. Inputs to the perceptron are fed either from an external source or by a nearby perceptron. The connection or links within the perceptron, known as weights, are similar to synapses in the biological neuron. The inputs are accumulated (Figure 5.2) and summed within the processing unit. The perceptron then makes a check as to whether the weighted

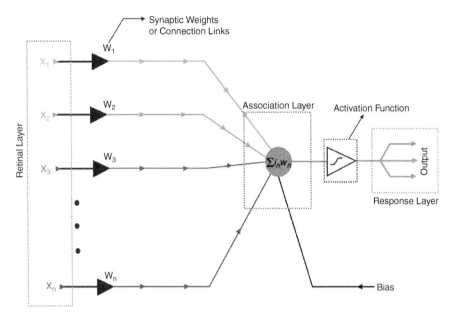

Figure 5.2   Typical architecture of the Mathematical Neuron (after Rosenblatt, 1958; Van der Baan & Jutten, 2000).

sum lies above or below a threshold value (Poulton, 2001), and accordingly the perceptron generates an output. Further, an error is computed between the perceptron-generated output and the desired output. The connection weights within the perceptron can be updated when the calculated and desired results do not match satisfactorily until the difference between these two cannot be reduced further. The mathematical neuron can be suitably designed to obtain an optimum output and exhibit a higher resemblance to the biological neuron.

Within the processing unit of a perceptron, the accumulated input signals undergo summation processes. Let us assume that the numbers of inputs, fed into the perceptron, are $x_1, x_2, x_3, x_4, x_5, \ldots x_i$ and the connection weights are $w_1, w_2, w_3, w_4, w_5, \ldots w_i$. The summation operation (y) is given by:

$$y = \sum x_i w_i. \tag{5.1}$$

This results in a linear summation process. However, a decisional output is always expected from the perceptron, for which this weighted sum should flow through a decisional unit. This comprises a non-linear transfer function (or an activation function) where the weighted sum of inputs is fed to rescale the sum. Now Eq. 5-1 can be rewritten as:

$$y = a\left(\sum x_i w_i\right), \tag{5.2}$$

where a is the activation or transfer function. A constant bias (θ) is applied to shift the position of the activation function, independent of the input signal (Van der Baan & Jutten, 2000). The bias is often described in many studies as the threshold. The above equation can be further modified as:

$$y = a\left(\sum x_i w_i - \theta\right). \tag{5.3}$$

5.4. Activation or Transfer Function

Earlier developments of the artificial perceptron employed the Heaviside or hard limiting function that resulted in binary outputs, not continuously differentiable ones (Van der Baan & Jutten, 2001). Thus, it was challenging to adjust the connection weights. A plausible solution was given by introducing the most widely used sigmoid function (Figure 5.3), which is continuously differentiable, monotonically increasing, and has a smoothed step function, and is expressed as:

$$f(a) = (1 + e^{-a})^{-1}. \tag{5.4}$$

Different non-linear activation functions, e.g., sigmoid, bipolar sigmoid, tangent hyperbolic, etc., are used for designing the mathematical perceptron. The

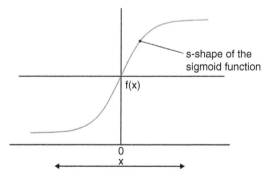

Figure 5.3 A typical s-shaped sigmoidal function.

non-linearity nature helps to generate a complex partition at the decision boundaries in the feature space and obtain discrete results for a perceptron in terms of any general and highly non-linear function from a given set of input variables (Aminzadeh & de Groot, 2006).

5.5. Types of Learning

Learning is a process adopted to train the perceptron. There exist three different learning mechanisms: *supervised learning, unsupervised learning,* and *reinforcement learning,* by which a perceptron can be trained to solve a specific job. The main difference between these learning techniques lies in the amount of a-priori information that is supplied to the perceptron (Aminzadeh & de Groot, 2006; Poulton, 2001).

A *supervised learning* approach is a process in which the perceptron or network is made to learn from available examples, i.e. one must provide correct values for each input pattern. The values assigned are then compared with the output generated by the network. In this comparison, the network attempts to minimize the error between the known and predicted output, and tries to establish a non-linear relationship between the input and target variables. *Reinforcement learning* is a type of supervised learning scheme where the only supplied information is the answer indicating whether the computed output is correct or incorrect. A reward is given for a correct answer and a penalty for the wrong answer. An *unsupervised learning* approach aims to identify the structure, and the system learns itself regarding the pattern present within the data pairs. In such a learning process, no a-priori information is provided to the system. The perceptron itself analyses and determines common features (i.e., structures) present within these inputs. Examples of such learning are Self-Organizing Maps (SOM) (Kohonen, 2001).

5.6. Multi-Layer Perceptron (MLP) and the Backpropagation Algorithm

As the name suggests, the perceptron comprises multi or different layers in which the nodes or neurons in each layer are fully connected to the nodes or neurons of the consecutive layers. It is also called a feedforward multi-layer perceptron that employs a backpropagation learning algorithm. Many authors simplify its name to backpropagation network because that is what solves the problem. Such a network is robust and widely used. This method was invented by Werbos (1974) and later reworked by Parker (1985). Further, the technique was rationalized through the pioneering works of McClelland et al. (1986), Sukhan (1988), and Rumelhart et al. (1995).

The MLP consists of three distinct layers: the input layer, the hidden layer, and the output layer. The input layer contains several inputs that are planned to be fed into the network processing elements (PEs). The neurons within this layer are not engaged in signal modification but are responsible for transmitting the signal into the next layer. The hidden layer consists of neurons that are responsible for accumulating, processing, and modifying incoming signals. The output layer contains a fixed number of outputs. The number of input and output patterns are decided from the tasks to be addressed. This is a supervised learning process, where the known examples comprising of input and output patterns are fed to the network during the training phase. The network learns and generates an output in a feedforward process (i.e., where the information flows in the forward direction). Once this is done, a *backpropagation learning* algorithm (LeCun, 1885; McClelland et al., 1986; Parker, 1985; Rumelhart et al., 1995; Werbos, 1974) is used to minimize the error between the predicted or generated output of the network and the desired or known output fed to the network by adjusting or updating the connection links or weights.

Let us look at the mathematical insights to understand the processing in the MLP system. Assume that each of the input, hidden, and output layers of an MLP system (Figure 5.4) is made up of 2 neurons (i_1 and i_2 for input layer; h_1 and h_2 for

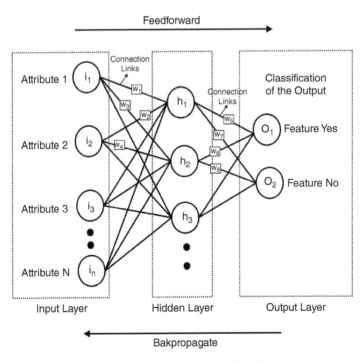

Figure 5.4   The multi-layer perceptron (MLP) network architecture, consisting of fully connected neurons within the input, hidden, and output layers.

hidden layer; o_1 and o_2 for the output layer), which are fully connected with connection weights w_i for i ranging from 1 to 8. Now, the value received at each of the processing elements (h_1 and h_2) in the hidden layer is given as:

$$h_1 = (i_1 \times w_1) + (i_2 \times w_2) = h_{net1} \tag{5.5}$$

$$h_2 = (i_1 \times w_3) + (i_2 \times w_4) = h_{net2}. \tag{5.6}$$

These are the outputs for the processing elements (h_1 and h_2) of the hidden layer. This output is then passed through an activation function—say, sigmoid function for the present case. The final output at h_1 becomes:

$$h_{out1} = f(h_{net1}), \tag{5.7}$$

where

$$f(h_{net1}) = \frac{1}{1 + e^{-h_{net1}}}. \tag{5.8}$$

Similarly, the final output at h_2 becomes:

$$h_{out2} = f(h_{net2}), \tag{5.9}$$

where

$$f(h_{net2}) = \frac{1}{1 + e^{-h_{net2}}}. \tag{5.10}$$

The signal received at the output layer is given by:

$$o_{net1} = (h_{out1} \times w_5) + (h_{out2} \times w_6)$$

$$o_{net2} = (h_{out1} \times w_7) + (h_{out2} \times w_8)$$

The final values at the output layer now become

$$o_{out1} = f(o_{net1}) \tag{5.11}$$

$$o_{out2} = f(o_{net2}). \tag{5.12}$$

The activation function modulates the connection weights as the signal flows from one layer to the next layer until it finally reaches the output layer. Here the output layer consists of two neurons, which in turn results in two outputs at each neuron. These are called the targeted and the computed outputs. When a mismatch between these outputs arises, the network applies some error corrections. This happens because we start the learning process by assigning some random values of weights. Let us designate the error encountered at o_1 as E_{o_1} and at o_2 as E_{o_1}, respectively. The error is given by:

$$E_{o_1} = \frac{1}{2} \left(Target\ o_{out\ 1} - Computed\ o_{out\ 1} \right)^2 \tag{5.13}$$

$$E_{o_2} = \tfrac{1}{2}\,(Target\ o_{out\ 2} - Computed\ o_{out\ 2})^2. \tag{5.14}$$

These errors are the mean-square error (MSE), and the square root of this error is called the RMS error. The total error at the neurons in the output layer is:

$$E_{total} = E_{o_1} + E_{o_2} \tag{5.15}$$

or

$$E_{total} = \tfrac{1}{2}\left(\begin{array}{c} (Target\ o_{out1} - Computed\ o_{out1})^2 \\ + (Target\ o_{out2} - Computed\ o_{out2})^2 \end{array} \right). \tag{5.16}$$

The primary goal is to minimize the error for the connection weights. This becomes possible when we take the derivative of the above equation. In the above equation the constant $\tfrac{1}{2}$ helps in canceling out the derivative. Let us do the exercise for the weight w_5.

By taking the partial derivatives and following the chain rule technique, we have:

$$\frac{\partial E_{total}}{\partial w_5} = \frac{\partial E_{total}}{\partial o_{out1}} \times \frac{\partial o_{out1}}{\partial o_{net1}} \times \frac{\partial o_{net1}}{\partial w_5}. \tag{5.17}$$

Let us solve individually the three terms on the right-hand side (RHS) of Eq. 5-17:

Solving for the first term we have

$$\frac{\partial E_{total}}{\partial o_{out1}} = \frac{\partial}{\partial o_{out1}}(E_{o_1} + E_{o_2}). \tag{5.18}$$

This equation finally becomes

$$\frac{\partial E_{total}}{\partial o_{out1}} = 2 \times \frac{1}{2}\,(Target o_{out1} - Computed o_{out1}) \tag{5.19}$$

or

$$\frac{\partial E_{total}}{\partial o_{out1}} = (Target o_{out1} - Computed o_{out1}). \tag{5.20}$$

Solving for the second term we have

$$\frac{\partial o_{out1}}{\partial o_{net1}} = \frac{\partial}{\partial o_{net1}}\left(\frac{1}{1 + e^{-o_{net1}}} \right) \tag{5.21}$$

or

$$\frac{\partial o_{out1}}{\partial o_{net1}} = \frac{\partial}{\partial o_{net1}}\left(\frac{1}{1 + e^{-o_{net1}}} \right).$$

Solving for the RHS of this equation,

$$\frac{\partial}{\partial o_{net1}} \left(\frac{1}{1 + e^{-o_{net1}}} \right) = \frac{0 - 1 \times (-e^{-o_{net1}})}{(1 + e^{-o_{net1}})^2}.$$

Adding and subtracting 1 in the numerator of the term in RHS,

$$\frac{\partial}{\partial o_{net1}} \left(\frac{1}{1 + e^{-o_{net1}}} \right) = \frac{1 + e^{-o_{net1}} - 1}{(1 + e^{-o_{net1}})^2}.$$

Again, rearranging the terms,

$$\frac{\partial}{\partial o_{net1}} \left(\frac{1}{1 + e^{-o_{net1}}} \right) = \frac{(1 + e^{-o_{net1}}) - 1}{(1 + e^{-o_{net1}})^2}$$

or

$$\frac{\partial}{\partial o_{net1}} \left(\frac{1}{1 + e^{-o_{net1}}} \right) = \frac{1}{(1 + e^{-o_{net1}})} - \frac{1}{(1 + e^{-o_{net1}})^2}$$

or

$$\frac{\partial}{\partial o_{net1}} \left(\frac{1}{1 + e^{-o_{net1}}} \right) = o_{out1} - (o_{out1})^2$$

or

$$\frac{\partial o_{out1}}{\partial o_{net1}} = o_{out1} (1 - o_{out1}). \tag{5.22}$$

Solving for the third term we have

$$\frac{\partial o_{net1}}{\partial w_5} = \frac{\partial}{\partial w_5} ((h_{out1} \times w_5) + (h_{out2} \times w_7)) \tag{5.23}$$

or

$$\frac{\partial o_{net1}}{\partial w_5} = (h_{out1}).$$

Thus, we can calculate the error gradient $\frac{\partial E_{total}}{\partial w_5}$. Once this is done, we need a learning rule to change the weights in proportion (η) to this gradient. The new connection weight (w_5^{new}) between the hidden and output layer is given as

$$w_5^{new} = w_5 - \eta \frac{\partial E_{total}}{\partial w_5} \tag{5.24}$$

or

$$(w_5^{new} - w_5) = \Delta w \approx -\eta \frac{\partial E_{total}}{\partial w_5}. \tag{5.25}$$

In this equation, the proportionality (η) is called the learning rate. The equation can further be rewritten by introducing the momentum (α):

$$\left(w_5^{new} - w_5\right) = \Delta w \approx -\eta \frac{\partial E_{total}}{\partial w_5} + \alpha w_5. \tag{5.26}$$

This equation demonstrates the weight adjustment required on a connection link (w_5) made between the hidden neuron h_1 and the output neuron o_1. Similar operations can be performed for the connection links, e.g., w_6, w_7, w_8, w_9 between the hidden and output layers and for the connection links, e.g., w_1, w_2, w_3, w_4 between the hidden and input layers.

Learning rate (η) and momentum (α) control the amount of change that needs to be incorporated into the connection links or weights. Learning rate and momentum values are very often assigned to the network on a trial-and-error basis. A small value of the learning rate slows down the convergence but ensures the global minimum. However, a large value of learning rate becomes suitable when the error surface is relatively flat (Poulton, 2001). The number of hidden layers required to solve a particular problem by the perceptron is often a matter of concern. Many researchers (Cybenko, 1989; Hecht-Nielsen, 1990; Hornik et al., 1989) have addressed this problem through several proofs. Bishop (1995) provided a practical suggestion that a network with one hidden layer using a sigmoidal activation function can be approximated as a continuous function with a sufficient number of processing elements in the layer. The inference drawn by most of the researchers is that one hidden layer is sufficient but the addition of another increases the accuracy and decreases the learning time.

5.7. Different Types of ANNs

5.7.1. Radial Basis Function (RBF) Network

The *radial basis function* neural network works on the philosophy that if one tries to map the supplied input patterns into a higher dimensional space, there exists a probability such that the problem becomes linearly separable in accordance with Cover's Theorem (Cover, 1965; Haykin, 1999; Poulton, 2001). The structure of this network possesses similarity with the MLP network but differs in the weight distribution. In the RBF network, weights are assigned only between the hidden layer and the output layer (Figure 5.5). Each neuron in the hidden layer is associated with a basis function, and the output is obtained through a summation process, given as the sum of n basis functions, each multiplied with the weights. Powell (1987) and Poggio and Girosi (1990) demonstrated an extensive application of the basis function in modeling of data. The reader may look at the pioneering works of Broomhead and Lowe (1988), Moody and Darken (1988), Platt (1991), and Aminzadeh and de Groot (2004).

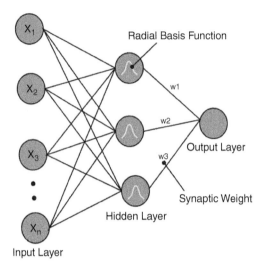

Figure 5.5 The RBF network architecture. Unlike in MLP (Figure 5.4), weights are assigned only to the hidden layer. However, all the neurons of each layer are interconnected to each other.

5.7.2. Probabilistic Neural Network (PNN)

The origin of the PNN dates back to 1960, and it evolved from the works of Specht (1991) at Stanford University. The PNN resembles the backpropagation network but there is a difference in which the sigmoidal threshold function is replaced with a probability density function (PDF) derived using Bayesian statistics. In backpropagation networks, the training is initialized with a random set of weights. However, in the PNN system, the network is initialized with equal weights to all the input patterns. A PNN architecture that consists of four different layers—the input layer, the pattern layer, the summation layer, and the output layer—is demonstrated in Figure 5.6. The processing elements in the pattern layer are not fully connected to the summation layer. However, these processing elements have connection weights to a particular input pattern, and are connected to the summation unit corresponding to that pattern's output class (Poulton, 2001). The training of the PNN is carried out using a competitive learning rule, according to which one pattern processing element is allowed to be active at a time. The processing elements in the summation unit adds up all the values for each class from the pattern layer. The element with maximum response in the summation layer shall result in output as classification. The PNN demonstrates its usefulness in solving supervised classification problems and trains the data very quickly. However, it is about 60 to 100 times slower than the backpropagation

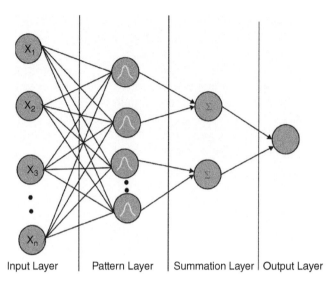

Input Layer | Pattern Layer | Summation Layer | Output Layer

Figure 5.6   The architecture of PNN demonstrating four different layers (input, pattern, summation, and output) interconnected to each other.

when used in recall mode (Poulton, 2001). To learn more about the practical applications of PNN, the reader may go through the works of Hampson et al. (2001) and Russell (2004).

5.7.3. Generalized Regression Neural Network (GRNN)

The GRNN (Specht, 1991) is a generalized form of the PNN and is used to address problems related to both estimation and classification. The network follows a linear regression approach and requires one hidden node for each training sample. To prevent network complexities, Specht (1991) clustered the training data so that a particular node in the data can respond to multiple input patterns. For more insights on the geophysical applications of the GRNN, readers may look into the works of Hampson & Todorov (1999).

5.7.4. Modular Neural Network (MNN)

The MNN consists of a group of individual neural networks, called local experts, which are interconnected to each other. These are also referred to as dynamic committed machines (Aminzadeh & de Groot, 2006; Poulton, 2001). The MNN holds the basic MLP architecture consisting of input and output layers. Several local experts reside within these layers, each of which is fully connected to the input layer. The output of each of these local experts is governed by a global expert known as the "gating network." This gating network has the ability to decide which local expert estimates an output closer to the desired value. Such a local expert becomes the winning expert of the network, whose connection weight is further updated and strengthened. The MNN thus segments the supplied training data into regions with similar patterns. The application of the MNN is explored in the works of Zhang and Poulton (2003) and Fruhwirth and Steinlechner (2004).

5.7.5. Self-Organizing Maps (SOM)

The SOM approach owes its development to the works of Kohonen (2001). The SOM network maps the input patterns that are spatially close to each other into a topographic map of the input (Figure 5.7) that demonstrates a natural relationship between the input patterns (Poulton, 2001). Let us assume an input pattern $\overline{x} = x_1, x_2, x_3, x_4, \ldots\ldots, x_n$ and let the weights assigned to individual units in the input pattern be $w_{ji} = [\ w_{j1}, w_{j2}, w_{j3}, \ldots\ldots\ldots, w_{jn}]$, where j is the processing elements within the layer, known as the Kohonen layer, and i belongs to the processing elements of the input layer. For each Kohonen processing element, the distance between the input pattern and the weight vector is computed. A matching value is determined that signifies the closeness of the weights in each unit into the corresponding input pattern. The match is given as:

$$Match_j = \sqrt{\sum\nolimits_{i=1}^{n} (x_i - w_{ji})^2}. \tag{5.27}$$

The unit possessing the lowest matching value is awarded in the competition. The neighborhood surrounding the winning unit in the Kohonen layer (or the

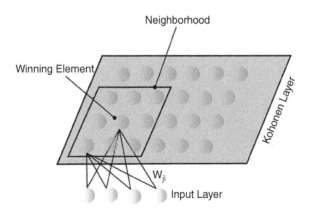

Figure 5.7   The SOM network in which the input layer is fully connected with the Kohonen layer. The weights are updated in the neighborhood of the winning element (after Poulton, 2002).

competitive layer) is then identified, and the weights of the processing elements within this neighborhood region are updated. The change in weight, or the delta weight, is given as:

$$\delta_{ji} = \eta \left(x_i - w_{ji} \right). \tag{5.28}$$

The weights are modified by adding this delta value to the old weights for the processing element within the winning neighborhood region. However, the weight adjustments for other processing elements within this layer are considered to be zero. The SOM mapping approach follows the scheme of the unsupervised process of learning, where the Kohonen layer identifies the structural pattern present in the data. The SOM technique, if grouped together with the backpropagation algorithm, can also be operated in the supervised mode of learning (Castellano & Fanelli, 2000; Poulton, 2001). Examples of SOM networks are available in several works (Carr et al., 2001; Himberg, 1998; Kaski & Kohonen, 1998; Kohonen, 2001; Sliwa et al., 2003; Strecker & Uden, 2002; Taner et al., 2001).

Some other networks or alternative architectures, which are not discussed in this book, are Hopfield Networks (HN) (Hopfield, 1984), Learning Vector Quantizer (LVQ) (Haykin, 1999; Hertz et al., 1991), Uniform Vector Quantizer (UVQ) (Aminzadeh & de Groot, 2006), etc., and they can also be used for addressing complex problems. The development of Deep Learning architecture such as convolutional neural networks (CNNs) has made a breakthrough in the application of machine learning techniques to handle very complicated real problems.

5.8. Summary

We have presented in this chapter a brief overview of ANNs, including the historical evolution of NNs and different network architectures that can be designed to address a variety of scientific problems. The next chapters explore how these networks could be used with geophysical field data for advanced interpretation of subsurface features, mainly 3D seismic data.

References

Aminzadeh, F., de Groot, P., & Wilkinson, D. (2004) Soft Computing for qualitative and quantitative seismic object and reservoir property prediction, Part 1, Neural network applications, Part 2, Fuzzy logic applications, Part 3, Evolutionary computing and other aspects of soft computing. *First Break*, vol. 22, pp. 49–54, pp. 69–78, pp. 107–116. https://doi.org/10.3997/1365-2397.22.6.25903

Aminzadeh, F., & de Groot, P. (2006). *Neural networks and other soft computing techniques with applications in the oil industry*. EAGE Publications.

Bishop, C. (1995). *Neural networks for pattern recognition*. Oxford Press.

Broomhead, D. S., & Lowe, D. (1988). *Radial basis functions, multi-variable functional interpolation and adaptive networks* (No. RSRE-MEMO-4148). Royal Signals and Radar Establishment Malvern (United Kingdom).

Caianiello, E. R. (1961). Outline of a theory of thought-processes and thinking machines. *Journal of theoretical biology*, *1*(2), 204–235. https://doi.org/10.1016/0022-5193(61) 90046-7

Calderón-Macías, C., Sen, M. K., & Stoffa, P. L. (2001). Artificial neural networks for parameter estimation in geophysics. *Geophysical prospecting*, *48*(1), 21–47. https://doi.org/10.1046/j.1365-2478.2000.00171.x

Calderón-Macías, C., Sen, M. K., & Stoffa, P. L. (1997). Hopfield neural networks, and mean field annealing for seismic deconvolution and multiple attenuation. *Geophysics*, *62*(3), 992–1002. https://doi.org/10.1190/1.1444205

Calderón-Macías, C. (1999). Artificial neural systems for interpretation and inversion of seismic data. *Geophysical Prospecting*, *48*(1), 21–47. http://dx.doi.org/10.1046/j.1365-2478.2000.00171.x

Carr, M., Cooper, R., Smith, M., Taner, M. T., & Taylor, G. (2001). The generation of a rock and fluid properties volume via the integration of multiple seismic attributes and log data. In *7th International Congress of the Brazilian Geophysical Society*, European Association of Geoscientists & Engineers, cp-217-00008. https://doi.org/10.3997/2214-4609-pdb.217.008

Castellano, G., & Fanelli, A. M. (2000). A self-organizing neural fuzzy inference network. In Proceedings of the IEEE-INNS-ENNS International Joint Conference on Neural Networks. IJCNN 2000. *Neural computing: New challenges and perspectives for the new millennium* (Vol. 5, pp. 14–19). IEEE. https://doi.org/10.1109/IJCNN.2000.861428

Chentouf, R., Jutten, C., Maignan, M., & Kanevsky, M. (1997). Incremental neural networks for function approximation. *Nuclear Instruments and Methods in Physics Research Section A: Accelerators, Spectrometers, Detectors and Associated Equipment*, *389*(1–2), 268–270. https://doi.org/10.1016/S0168-9002(97)00081-8

Cover, T. M. (1965). Geometrical and statistical properties of systems of linear inequalities with applications in pattern recognition. *IEEE Transactions on Electronic Computers*, *3*, 326–334. https://doi.org/10.1109/PGEC.1965.264137

Cybenko, G. (1989). Approximation by superpositions of sigmoidal functions. *Mathematics of Control, Signals and Systems*, *2*, 303–314. https://doi.org/10.1007/BF02551274

Fischbach, G. (1992). Mind and brain. *Scientific American*, *267*(3), 48–59.

Fruhwirth, R. K., & Steinlechner, S. P. (2004). A systematic approach to the optimal design of feed forward neural networks applied to log-synthesis. In EAGE 66th Conference & Exhibition.

Hampson, D., & Todorov, T. (1999). AVO lithology prediction using multiple seismic attributes. Paper presented in *SEG Technical Program Expanded Abstracts* (pp. 756–759). Society of Exploration Geophysicists.

Hampson, D. P., Schuelke, J. S., & Quirein, J. A. (2001). Use of multiattribute transforms to predict log properties from seismic data. *Geophysics*, *66*(1), 220–236. https://doi.org/10.1190/1.1444899

Haykin, S. (1999). *Neural networks: A comprehensive foundation* (2nd ed.). Prentice Hall.

Hebb, Donald O. (1949). *The organization behavior*. Wiley

Hecht-Nielsen, R. (1990) *Neurocomputing*. Addison-Wesley.

Hertz, J., Krogh, A., & Palmer, R (1991). *Introduction to the theory of neural computation*. Addison Wiley. https://doi.org/10.1201/9780429499661

Himberg, J. (1998). Enhancing the SOM-based data visualization by linking different data projections. In *Proceedings of 1st International Symposium IDEAL* (Vol. *98*, pp. 427–434).

Hopfield, J. J. (1984). Neurons with graded response have collective computational properties like those of two-state neurons. *Proceedings of the National Academy of Sciences, 81* (10), 3088–3092. https://doi.org/10.1073/pnas.81.10.3088

Hornik, K., Stinchcombe, M., & White, H. (1989). Multilayer feedforward networks are universal approximators. *Neural Networks, 2*(5), 359–366. https://doi.org/10.1016/0893-6080(89)90020-8

Huang, Z., Shimeld, J., Williamson, M., & Katsube, J. (1996). Permeability prediction with artificial neural network modeling in the Venture gas field, offshore eastern Canada. *Geophysics, 61*(2), 422–436. https://doi.org/10.1190/1.1443970

Hush, D. R., & Horne, B. G. (1993). Progress in supervised neural networks. *IEEE signal processing magazine, 10*(1), 8–39. https://doi.org/10.1109/79.180705

Kaski, S., & Kohonen, T. (1998). Tips for processing and color-coding of self-organizing maps. In G. Deboeck & T. Kohonen (Eds.), *Visual explorations in finance with self-organizing maps*, pages 195–202. Springer, London. https://doi.org/10.1007/978-1-4471-3913-3_14

Kohonen, T. (2001). *Evolution of ideas for self-organizing neural networks*. Paper presented in Asian Pacific Symposium on Life Science and Systems Engineering, 25–26 July, 8–36.

Kumar, P. C., & Mandal, A. (2017). Enhancement of fault interpretation using multi-attribute analysis and artificial neural network (ANN) approach: A case study from Taranaki Basin, New Zealand. *Exploration Geophysics, 49*(3), 409–424. https://doi.org/10.1071/EG16072

Kumar, P. C., & Sain, K. (2018). Attribute amalgamation-aiding interpretation of faults from seismic data: An example from Waitara 3D prospect in Taranaki basin off New Zealand. *Journal of Applied Geophysics, 159*, 52–68. https://doi.org/10.1016/j.jappgeo.2018.07.023

Kumar, P. C., Omosanya, K. O., & Sain, K. (2019a). Sill Cube: An automated approach for the interpretation of magmatic sill complexes on seismic reflection data. *Marine and Petroleum Geology, 100*, 60–84. https://doi.org/10.1016/j.marpetgeo.2018.10.054

Kumar, P. C., Sain, K., & Mandal, A. (2019b). Delineation of a buried volcanic system in Kora prospect off New Zealand using artificial neural networks and its implications. *Journal of Applied Geophysics, 161*, 56–75. https://doi.org/10.1016/j.jappgeo.2018.12.008

Kumar, P. C., Omosanya, K. O., Alves, T. & Sain, K. (2019c). A neural network approach for elucidating fluid leakage along hard-linked normal faults. *Journal of Marine and Petroleum Geology, 110*, 518–538. https://doi.org/10.1016/j.marpetgeo.2019.07.042

Kumar, P. C., & Sain, K. (2020). Interpretation of magma transport through saucer sills in shallow sedimentary strata using an automated machine learning approach. *Tectonophysics, 789*, 228541, 1–16. https://doi.org/10.1016/j.tecto.2020.228541

Langer, H., Nunnari, G., & Occhipinti, L. (1996). Estimation of seismic waveform governing parameters with neural networks. *Journal of Geophysical Research: Solid Earth, 101* (B9), 20109–20118. https://doi.org/10.1029/96JB00948

Le Cun, Y. 1985. A learning procedure for asymmetric network. *Proc. Cognit. (Paris), 85*, 599–604.

Moody, J., & Darken, C. (1988). *Learning with localized receptive fields* (pp. 133–143). Yale Univ., Department of Computer Science.

McClelland, J. L., Rumelhart, D. E., & PDP Research Group. (1986). Parallel distributed processing. *Explorations in the Microstructure of Cognition, 2*, 216–271.

McCormack, M. P. (1991). Neural networks in the petroleum industry. In *SEG Technical Program Expanded Abstracts 1991* (pp. 728–731). Society of Exploration Geophysicists. https://doi.org/10.1190/1.1889172

McCormack, M. D., Zaucha, D.E., & Dushek, D.W. (1993). First-break refraction event picking and seismic data trace editing using neural networks. *Geophysics*, *58*(1): 67–78. https://doi.org/10.1190/1.1443352

McCulloch, W. S., & Pitts, W. (1943). A logical calculus of the ideas immanent in nervous activity. *The bulletin of mathematical biophysics*, *5*(4), 115–133. https://doi.org/10.1007/BF02459570

Murat, M. E., & Rudman, A.J. (1992). Automated first arrival picking: A neural network approach 1. *Geophysical Prospecting*, *40*(6), 587–604. https://doi.org/10.1111/j.1365-2478.1992.tb00543.x

Nikravesh, M., Aminzadeh, F., & Zadeh, L.A. (2003). *Soft computing and intelligent data analysis*. Elsevier.

Parker, D. (1985). Learning-logic: Technical Report TR-47, Center for Computational Research in Economics and Management Science, MIT, April

Platt, J. C. (1991). Leaning by combining memorization and gradient descent. In *Advances in Neural Information Processing Systems* (pp. 714–720).

Poggio, T., & Girosi, F. (1990). Networks for approximation and learning. *Proceedings of the IEEE*, *78*(9), 1481–1497.

Poulton, M. M., Sternberg, B. K., & Glass, C. E. (1992). Location of subsurface targets in geophysical data using neural networks. *Geophysics*, *57*(12), 1534–1544. https://doi.org/10.1190/1.1443221

Poulton, M. M. (Ed.) (2001). *Computational neural networks for geophysical data processing*. Elsevier, Amsterdam, The Netherlands.

Poulton, M. M. (2002). Neural networks as an intelligence amplification tool: A review of applications. *Geophysics*, *67*(3), 979–993. https://doi.org/10.1190/1.1484539

Powell, M. J. (1987). Radial basis functions for multivariable interpolation: a review. *Algorithms for approximation*.

Rochester, N., Holland, J., Haibt, L., & Duda, W. (1956). Tests on a cell assembly theory of the action of the brain, using a large digital computer. *IRE Transactions on information Theory*, *2*(3), 80–93. https://doi.org/10.1109/TIT.1956.1056810

Röth, G., & Tarantola, A. (1994). Neural networks and inversion of seismic data. *Journal of Geophysical Research: Solid Earth*, *99*(B4), 6753–6768. https://doi.org/10.1029/93JB01563

Rosenblatt, F. (1958). The perceptron: a probabilistic model for information storage and organization in the brain. *Psychological Review*, *65*(6), 386–408. https://doi.org/10.1037/h0042519

Russell, B. H. (2004). The application of multivariate statistics and neural networks to the prediction of reservoir parameters using seismic attributes. *University of Calgary 6255*, 1–16.

Sain, K. & Kumar, P. C. (2021). Seismic, neural intelligence to artificial intelligence for seismic interpretation. In H.K. Gupta (Ed.), *Encyclopedia of Solid Earth Geophysics* (2nd Edition). Springer.

Sandham, W., Leggett, L., & Aminzadeh. F. (2003). *Applications of artificial neural networks and fuzzy logic*. Kluwer Academic Publisher. Introduction.

Sliwa, R., Fraser, S. J., & Dickson, B. L. (2003). Application of self-organising maps to the recognition of specific lithologies from borehole geophysics. In *Proceedings of the 35th Sydney Basin Symposium on "Advances in the study of the Sydney Basin"* (pp. 105–113).

Singh, D., Kumar, P. C., & Sain, K. (2016). Interpretation of gas chimney from seismic data using artificial neural network: A study from Maari 3D prospect in the Taranaki basin, New Zealand. *Journal of Natural Gas Science and Engineering, 36*, 339–357. https://doi.org/10.1016/j.jngse.2016.10.039

Specht, D. F. (1991). A general regression neural network. *IEEE Transactions on Neural Networks, 2*(6), 568–576.

Strecker, U., & Uden, R. (2002). Data mining of 3D poststack seismic attribute volumes using Kohonen self-organizing maps. *The Leading Edge, 21*(10), 1032–1037. https://doi.org/10.1190/1.1518442

Sukhan, L. (1988). *Multilayer feedforward potential function network.* Paper presented in IEEE 1988 International Conference on Neural Networks (pp. 161–171). https://doi.org/10.1109/ICNN.1988.23844

Taner, M. T., Berge, T., Walls, J. D., Smith, M., Taylor, G., Dumas, D., & Carr, M. B. (2001). Well log calibration of Kohonen-classified seismic attributes using Bayesian logic. *Journal of Petroleum Geology, 24*(4), 405–416. https://doi.org/10.1111/j.1747-5457.2001.tb00683.x

Tonn, R. (2002). Neural network seismic reservoir characterization in a heavy oil reservoir. *The Leading Edge, 21*(3), 309–312. https://doi.org/10.1190/1.1463783

Van der Baan, M., & Jutten, C. (2000). Neural networks in geophysical applications. *Geophysics, 65*(4), 1032–1047. https://doi.org/10.1190/1.1444797

Vaishya, R., Javaid, M., Khan, I. H., & Haleem, A. (2020). Artificial Intelligence (AI) applications for COVID-19 pandemic. *Diabetes & Metabolic Syndrome: Clinical Research & Reviews*, 337–339. https://doi.org/10.1016/j.dsx.2020.04.012

Wang, L. X., & Mendal, J. M. (1991). Adaptive minimum prediction-error deconvolution and wavelet estimation using Hopfield neural networks. In *Acoustics, Speech, and Signal Processing, IEEE International Conference on* (pp. 2969–2970). IEEE Computer Society. https://doi.org/10.1190/1.1443281

Wang, L. X., & Mendal, J. M. (1992). Adaptive minimum prediction-error deconvolution and source wavelet estimation using Hopfield neural networks. *Geophysics, 57*(5), 670–679. https://doi.org/10.1190/1.1443281

Werbos, P. (1974). Beyond regression: New tools for prediction and analysis in the behavioral sciences: PhD Dissertation, Applied Math, Harvard University, Cambridge, MA

Widrow, B., & Hoff, M. E. (1960). *Adaptive switching circuits* (No. TR-1553-1). Stanford Univ Ca Stanford Electronics Labs.

Wong, P. M., Aminzadeh, F., & Nikravesh, M. (Eds.) (2002). *Soft Computing for Reservoir Characterisation and Modeling, Studies in Fuzziness and Soft Computing.* Physica-Verlag, Springer-Verlag. https://doi.org/10.1007/978-3-7908-1807-9

Zhang, Y., & Paulson, K. V. (1997). Magnetotelluric inversion using regularized Hopfield neural networks. *Geophysical prospecting, 45*(5), 725–743. https://doi.org/10.1046/j.1365-2478.1997.660299.x

Zhang, L., & Poulton, M. (2003). Neural Network Inversion of EM39 induction log data. In *Geophysical applications of artificial neural networks and fuzzy logic* (pp. 231–249). Springer, Dordrecht. https://doi.org/10.1007/978-94-017-0271-3_15

6

HOW TO DESIGN META-ATTRIBUTES

Computation of a meta-attribute from a set of other seismic attributes is crucial for delimiting subsurface geologic features and
hence to accelerate the interpretation of seismic data. This chapter explains how to design workflows and train the machine for the
computation of meta-attributes that can be suited for delineating
different geologic bodies. Most of these are based on a widely
used network called the Multi-Layer Perceptron (MLP), through a
supervised mode of neural learning for the interpretation of subsurface features from 3D as well as 2D seismic data.

6.1. Introduction

In the previous chapters, we learned about meta-attributes and understood
the essence of merging attributes through artificial neural networks for robust
interpretation of targeted object/property from seismic data. This chapter
describes the designing of workflows that can be used for the computation of
meta-attributes. Most of these are done using the widely used network called
the multi-layer perceptron (MLP) through a supervised mode of neural learning
for interpretation of subsurface features from 3D seismic volume. This process
can also be pursued from 2D seismic data, and in that case, the meta-attribute cube
(say, Fault Cube) is simply referred to as meta-attribute (Fault) only. The MLP
process can even be employed in 1D data such as vertically drilled well data. Very
recently, well logs and available core data have been utilized for the identification
of lithological sequences with variable thickness based on the artificial neural

Meta-Attributes and Artificial Networking: A New Tool for Seismic Interpretation,
Special Publications 76, First Edition. Kalachand Sain and Priyadarshi Chinmoy Kumar.
© 2022 American Geophysical Union. Published 2022 by John Wiley & Sons, Inc.
DOI: 10.1002/9781119481874.ch06

approach (Mukherjee & Sain, 2019, 2021; Singh et al., 2020). This has demonstrated the usefulness of this approach in assessing the results over the traditional wireline log interpretation. Let us start step by step to build the workflows and hence the meta-attributes.

6.2. Meta-Attribute Design

The design of a meta-attribute involves five steps. The first step starts with cleaning the input data, i.e. the *Data Conditioning*, in which noisy events that obscure geologic features are suppressed and the signal/noise ratio is enhanced. The second step deals with *Data Preparation*, i.e. selection of appropriate seismic attributes that can capture responses of the targets or objects. The third step includes *Train and Test Data Preparation*, i.e. plausible example locations selected by the interpreter based on his/her acquaintances with the properties and characteristics of the targets. The fourth step is to train the neural network, i.e., *Neural Network Design* that aims for a minimum RMS error between the computed output and the assigned output. If this is not achieved in both the train and test data sets, the network is re-checked and re-run through the feedforward and backpropagation approaches. After obtaining the minimum RMS error in both the train and test data sets, the output is quality checked over key seismic lines. If satisfied, the process is automated in the fifth step by running the network over the entire seismic volume to obtain a *Meta-Attribute Cube*. Figure 6.1 demonstrates a simplified workflow adopted to design such a meta-attribute.

6.2.1. Seismic Data Conditioning

When the seismic signal is contaminated with unwanted components, defined as noise, interpretation of subsurface features becomes tedious. Presence of seismic noises (both random and coherent) may result in artefacts and hence hamper steady interpretation. No matter how well the robust processing steps are applied to the data, the processed seismic volumes may still be contaminated by noises (Chopra & Marfurt, 2007). Moreover, their presence might not hamper an interpreter in generating time structure maps but would result in noisy, poor quality attribute volumes that may impede healthy interpretation. Hence, several noise-suppression algorithms have been developed, such that the structural and stratigraphic contents are well preserved and become noise-free. There exist two different tools that can be applied to the migrated seismic data to aid interpretation.

The first tool is the structure-oriented filter (SOF), which is applied to the time-migrated seismic data to differentiate between the dip-azimuth of seismic

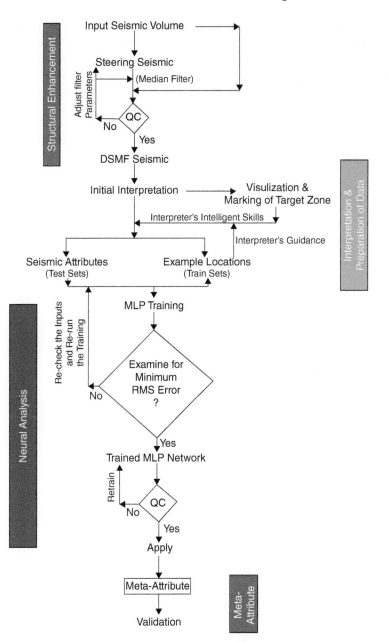

Figure 6.1 The generalized workflow used for designing meta-attribute.

reflectors and the overlying noise (Chopra & Marfurt, 2007). This process helps to remove random noises from the seismic data and enhance the lateral continuity of seismic events (Höcker & Fehmers, 2002), and ultimately provides a sharp definition of geological structures (Kumar & Mandal, 2017). The SOF task begins, once we estimate the dips and azimuths of reflectors from given seismic data. Tingdahl (1999) and Tingdahl and de Groot (2003) term such estimation computation of a steering volume, which contains dip and azimuth information stored at every sample location within the data volume. After estimating the dip-azimuth volume, we apply a filter to enhance the signal along with the reflection from seismic volume. The most widely used SOFs are mean, median, and alpha-trimmed mean filters. The SOF plays significant role in image enhancement that makes the data ready for extraction of seismic attributes and hence computation of the meta-attribute.

Mean Filter (or Running Average Filter)

The mean filter or running average filter also, called the low pass filter, is mostly used as a noise-suppression filter. The mean filter operates in a running window mode and provides sample mean at the center within the analysis window. It generally uses an odd-numbered window size, i.e. 3×3 or 5×5, and may be either rectangular or elliptical (Al-Shuhail et al., 2017; Chopra & Marfurt, 2007). Mathematically the mean filter is defined as:

$$d_{mean} = \frac{1}{j} \sum_{j=1}^{J} d_j. \tag{6.1}$$

Where dj denotes the jth trace falling within the analysis window at time t. In a case study in the Fort Worth Basin, Texas, Al-Dossary and Marfurt (2007) demonstrated that the mean filter significantly improves the quality of meaningful long-wavelength features. However, it obliterates the visibility of short-wavelength fracture patterns. A median filter, on the other hand, can preserve the edges of fault blocks and stratigraphic features. However, it smears the visibility of narrow curvilinear features associated with the joints and fractures (Al-Shuhail et al., 2017).

Median Filter

The median filter is also widely used for seismic interpretation. It is routinely used in Vertical Seismic Profiling (VSP) as a velocity filter for distinguishing between the downgoing and upcoming events using the difference in apparent velocities (Al-Shuhail et al., 2017). The median filter works by replacing each sample with the median of the samples falling within the analysis window. The filter operates in an odd-numbered window size, e.g., 3×3 or 5×5. If there are J samples within an analysis window, the median is computed by initially using an ordering of k applied to the J samples and then applying the median. The sequence is given as:

$$d_{j(1)} \leq d_{j(2)} \leq d_{j(3)} \cdots\cdots\cdots \leq d_{j(k+1)} \cdots\cdots \leq d_{j(J)}. \tag{6.2}$$

The median is expressed as:

$$d_{median} = d_{j[k = (J+1)/2]}. \tag{6.3}$$

The median filter works well for preserving sharp discontinuities and removing impulse noises from the signal (Al-Shuhail et al., 2017; Chopra & Marfurt, 2007; Kumar & Sain, 2018). Figure 6.2 shows the application of 3- and 5-point median filters.

Alpha-Trimmed Mean Filter

The alpha-trimmed mean filter operates between the mean and the median filter and is mathematically given as:

$$d_\alpha = \frac{1}{(1-2\alpha)J} \sum_{k = \alpha J + 1}^{(1-\alpha)J} d_{j(k)}, \tag{6.4}$$

where $0 \leq \alpha \leq 0.5$. For $\alpha = 0.5$, the above equation becomes a median filter and with $\alpha = 0.0$, the above equation squeezes to the conventional mean filter. Compared to the mean filter, the median and alpha-trim mean filters show better performance in suppressing the noise and preserving subsurface geological details (Al-Shuhail et al., 2017).

Several well-known interpretation platforms, e.g., Petrel™, SeisEarth™, Paleoscan™, Decisionspace™, OpendTect™, IHS-Kingdom™, etc. are available for interpretation of seismic data. The conditioning workflows differ from

(a)

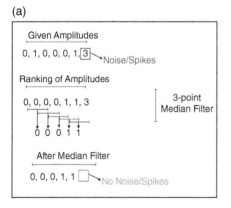

(b)

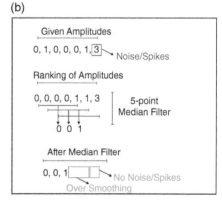

Figure 6.2   Operation of the median filter using a 3-point (a) and 5-point (b) filtering window. Note that both the filters efficiently remove the noise bursts. However, increasing the filter size results in over-smoothing and loses amplitude (b).

one platform to the other, but they make use of the above-mentioned basic filters and principles for cleaning the data and making it ready for interpretation.

Most of the studies described in this book are carried out using the OpendTect^{TM} interpretation platform. To condition the seismic data, a dip-azimuth filtering is applied to remove random noises that commonly obscure the subsurface features. This structural conditioning starts with dip-azimuth extraction of seismic reflectors using a technique known as dip-steering that results in a steering cube (or a dip-azimuth volume), which is achieved by using the phase-based dip algorithms (Jaglan et al., 2015; Kumar & Mandal, 2017; Tingdahl, 1999; Tingdahl & de Groot, 2003). This contains local dips of seismic reflectors and associated discontinuities. The steering cube is then passed through a median filter both locally and regionally to check the data quality. A mild filtering step-out, i.e., inl: xrl: sample: 1×1×3 is applied over the steering cube to understand the local variation. The output, called the detailed steering cube, stores detailed information of seismic reflectors. However, for defining the overall trend, coarser filtering is further applied by a filtering step-out, i.e., inl: xrl: sample 5×5×5. The resulting steering cube is called the background steering cube and contains overall dip trends of seismic reflectors and outlines background structural information. This steering cube is now taken as input for data enhancement through a process of structural filtering.

The structural filtering is performed using the SOFs. The filter operates on three basic principles: (i) orientation of seismic reflectors; (ii) identification of reflection terminations; and (iii) preservation of these terminations, thereby smoothing the reflectors. Based on these principles, the seismic volume is filtered in three different stages to obtain three different conditioned seismic volumes. These filtering stages are (i) smoothing the seismic reflectors through a dip-steered median filter (DSMF), (ii) enhancing the discontinuous positions through a dip-steered diffusion filter (DSDF), and (iii) logically merging these steps to generate fault-enhanced filtered (FEF) seismic data, which presents sharp and enhanced images of discontinuous structures. The DSMF is a statistical filter applied to the seismic volume using the pre-processed steering cube for obtaining a smoothed seismic volume where the continuity of seismic reflections is improved by suppressing random background noises (Jaglan et al., 2015). The filter applies median statistics over seismic amplitudes following the seismic dips. This is performed by applying a 5×5 median filtering step-out, and the output is the DSMF seismic data. In the next step, we move forward to improving the reflections closer to fault zones. The seismic data generally possess a diffusing character closer to these zones. To enhance the visibility of these zones, the DSDF, an intermediate filter, is used. The filter evaluates the quality of seismic data in a dip-steered circle. The central amplitude is replaced by the amplitude where the quality is deemed best (Jaglan et al., 2015; Kumar & Mandal, 2017), resulting in the enhancement of faults. The diffusion filter is guided by a pre-computed similarity attribute (computed using original stacked seismic and background steering data), as the presence of discontinuous structures is highlighted with low similar values. The final output of this operation is the DSDF seismic data. Finally, both the data volumes are logically merged (using a cut-off value of 0.5) with the help of a similarity attribute to obtain a Fault-Enhanced Filter (FEF), which performs the edge preserving and sharpening operation. The use of DSMF and FEF seismic volumes depends on the interpretation targets. For example, we feel for structural interpretation tasks the FEF seismic data is advantageous as it contains sharp, distinct edges of seismic reflectors, representing the discontinuous structures. However, the use of DSMF seismic volume is good enough for interpreting geologic targets such as gas clouds, sill networks, magmatic conduits, volcanic complex, carbonate reefs, mass transport deposits, channel complexes, etc.

6.2.2. Selection and Extraction of Seismic Attributes

In the first part of this book, we discussed several seismic attributes developed over time for the interpretation of geologic features. This is very apt when raising the question of which attributes should be used in generating a meta-attribute to best define a geologic target. A candid answer to such a question is that the selection of attributes is purely based on the observable responses from the geologic target of interest. The geologic targets are associated with certain seismic characteristics. For example, the discontinuous structures like geologic faults are generally associated with reflector terminations, vertical disturbances, and discontinuities in seismic reflectors (Chopra & Marfurt, 2007; Tingdahl, 2003; Tingdahl & de Groot, 2003). Faults are also characterized by abrupt changes in dips of reflectors that are generally oriented with different geometric shapes like curve, straight, and bend (Chopra & Marfurt, 2007; Roberts, 2001; Tingdahl, 1999, 2003; Tingdahl & de Groot, 2003; Tingdahl & de Rooij, 2005). Moreover, fault zones are usually associated with loss of signal energy and frequency content. From these clues, an interpreter can select geometric and physical seismic attributes, e.g., coherency, similarity, curvature, energy, frequency, etc., which can be blended into a fault meta-attribute. Assume another geologic target, such as the magmatic sill complex. Sills are concordant intrusive bodies, commonly sub-horizontal with a gentle inclination. They cross-cut the stratigraphy and exhibit upwards concave cross-sectional geometries with discordant limbs that help in transporting magmatic fluids into the overlying stratigraphic units. The study of sills is important in understanding their impact on the basin history and petroleum system (Cartwright & Møller Hansen, 2006; Omosanya, 2018; Smallwood & Maresh 2002; Thomson & Hutton 2004). They appear as saucer-shaped, transgressive, and gently inclined on seismic section. Magmatic sills exhibit high amplitude and entropy character due to larger impedance contrast with the surrounding sedimentary succession (Chopra & Marfurt, 2007). In such a scenario, reflection strength (or the amplitude envelope), texture entropy, and texture contrast would be good choices for generating a meta-attribute.

Thus, the selection of seismic attributes solely depends on seismic properties and characteristics associated with the geologic target and the interpreter's familiarities. Once these are selected, critical care must be taken to assign inline and xline step-outs and time gates (or window lengths), i.e., to parametrize the attributes. The selected attributes for amalgamation are generally extracted using three separate time windows: long, mid, and short (Aminzadeh & de Groot, 2006; Kumar & Sain, 2018; Tingdahl & de Rooij, 2005). For example, one can choose 64 ms or 96 ms for long, 32 ms or 48 ms for mid, and 16ms or 24ms for short time windows, respectively. The window lengths are designed to match the orientation and extension of targeted features throughout the data volume. Moreover, assigning accurate window lengths helps to improve the detection power of seismic attributes. The numerical values are not mandatory, but depend on the interpreter's choice. Once parametrized, the output is quality checked over vertical cross-sections and time slices to test the efficacy of imaging geologic structures. After being satisfied, the attributes are then grouped as test data sets.

6.2.3. Example Locations

Here human intelligence plays the role through which the network gets guidance in discriminating the character found within the input attributes from that of the fault and non-fault zones. This is done by selecting appropriate example locations (x, y, z) from the seismic volume, and such selection depends solely on the intuition of the interpreter. Assume that the target of interest/interpretation is the Mass Transport Deposit (MTD) from seismic data. In such a scenario, the interpreter classifies the seismic data into two categories i.e., the MTD zone and the non-MTD zone. For this, the interpreter randomly looks into seismic lines for the presence of MTD and non-MTD zones (Figure 6.3).

Ambiguous areas are those in which the interpreter is not sure about the target. It is observed that the MTD zone is characterized by disrupted and chaotic reflections and is internally deformed (Figures 6.3 and 6.4). Apart from the MTD, we also find several other geological features, e.g., faults, depositional beds, etc. Since our interest is to delineate the MTD, it should remain isolated from other zones for the success of attribute combination and delineation of geometry or architecture of the target. The result of the similarity attribute brings out all possible low similar values representing the faults, stratigraphic packages, and the MTD zone. Thus, the correct example location should be the MTD-yes zone and all other false locations should be marked as the MTD-no zone. A binary system of classification is adopted, where all MTD-yes zones correspond to 1, meaning the highest probability of MTD, and those of MTD-no correspond to 0, meaning the lowest or least likelihood of MTD.

Strictly speaking, the example location means the target zone and non-target zone within the given seismic data volume. The location consists of x,

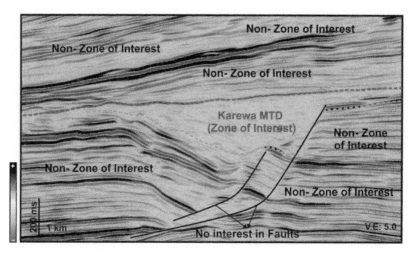

Figure 6.3   Seismic section displaying the MTD and non-MTD zones of interest from the Karewa prospect, offshore NZ. The top of the Karewa MTD is marked by the red dotted line, whereas the basal shear surface (BSS) is indicated by the green dotted line. The black solid lines represent the faults.

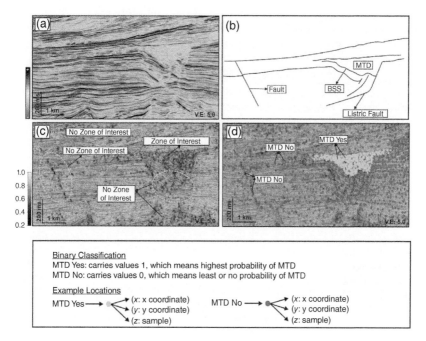

Figure 6.4   (a) Uninterpreted seismic section from Karewa 3D prospect. (b) Sketches of interpretation perceived after examining the seismic section in (a). (c) Seismic section co-rendered with computed similarity attribute demonstrating the zones of interests and non-interests. (d) Assigned example locations. The green dots refer to MTD-yes location and red dots refers to MTD-no location.

y, and z, which defines the position of the particular target and its extension. An interpreter can make the classifications not only over the vertical cross-sections but also on the time slices of the data. The main aim at this point is to create an even distribution of object and non-object example locations such that a true representative of the training locations is obtained from the data volume.

6.2.4. NN Operation

The NN learns through examples, gets trained, and establishes a relationship between input (seismic attributes) and output (object and non-object) sets. The training is performed using an MLP network that consists of input, hidden, and output nodes fully connected to each other. When the data flow between these layers, it is called the feedforward process. It is to be noted that a binary system is adopted, where all the object zones correspond to 1 with the highest probability of object, and all the non-object zones correspond to 0 with the lowest or least likelihood of object. Optimization of the network is carried out by a backpropagation algorithm (LeCun, 1985; Parker, 1985; Rosenblatt, 1962; Rumelhart et al., 1995; Werbos, 1974) attempting to minimize the error between predicted network result and known output by automatically adjusting the connection weights and other network parameters, e.g., learning rate and momentum. The network starts its operation by randomly splitting the data at the randomly selected nodes from a small volume (~20–30%) of entire data into train and test data sets. In general, the network is trained by taking 70% of data from the selected nodes or locations, while keeping remaining 30% data for testing. In other words, only 70% of the chosen data is used for training in which the related attributes are taken as input to compute the response lying between 0 and 1 using a feedforward process (Poulton, 2001, 2002; Rosenblatt, 1962). The network parameters are automatically adjusted iteratively based on backpropagation algorithm (LeCun, 1985; Parker, 1985; Poulton, 2001, 2002; Rosenblatt, 1962; Rumelhart et al., 1995; Werbos, 1974) to minimize the difference between the network response and the observation at training locations. Since the network also computes responses at the remaining 30% of locations (test data), the difference between the network response and the observation at the test locations is also calculated simultaneously to see if the neural model is trained correctly by observing the behavior, i.e. the decreasing trend of difference with iterations (Figure 6.5). Iterative neural training is continued until a minimum normalized RMS (nRMS) error and misclassification percentage (evaluation parameters used in this book) between the observed and computed response is achieved such that a probability output is obtained at all picked locations. At this stage, it is recommended to check the predictions over a few key seismic lines for quality check before applying the NN over the entire seismic volume to obtain a meta-attribute volume. The attribute volume contains samples whose values range from 0 to 1, where 0 signifies the least probability of the object and 1 signifies the highest certainty for the presence of the object. An optimum color scale is used in such a way that the maximum

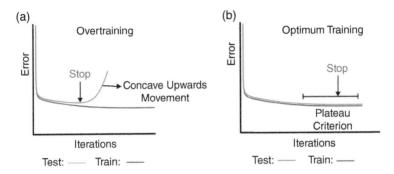

Figure 6.5 (a) Schematic representation of error curves demonstrating the overtraining situation. Neural training should be stopped when the test curve starts showing a concave upward pattern. (b) Schematic representation of error curves indicating the optimum training situation. This is achieved when both the error curves undergo a smooth decay and become a plateau with iterations.

probability, i.e. values closer to 1 are displayed and visualized, and those pertaining to values closer to 0, i.e. the least probability, are made transparent. The meta-attribute outputs, demonstrated in Chapters 7 to 14, are displayed using two different color scales, i.e. (a) white through grey to red and (b) pastel, by fixing the threshold value of 0.75 probability from the outcome.

Most importantly, the MLP network possesses two important features, e.g., abstraction (the ability to extract relevant features from the input pattern and discard the irrelevant ones) and generalization (recognize the input patterns which are not part of the training set). During the training phase, it is crucial to identify a proper stopping criterion in order to escape the overtraining that occurs when the NN finds non-universal relations at the example locations. Let us consider the interpretation target demonstrated in Section 6.2.3. Suppose, while assigning example locations to a few seismic lines in the cube, the interpreter cannot identify the correct MTD zone in the data but assigns MTD-yes there or considers as the zone of interest what is devoid of features. At this point, the training becomes non-universal, meaning the network will fail to judge whether these locations belong precisely to MTD or partially to MTD. Hence, the network is said to be over-trained, as it is unable to find a universal solution to such a perplexing situation. An optimum training (Figure 6.5) is generally obtained when the error within the train and test data sets reaches a plateau (Kumar & Sain, 2018; Kumar et al., 2019a, 2019b, 2019c).

Evaluation of Intelligent Neural Model

The performance evaluation of neural training in the present study is tracked through two different parameters, namely the nRMS and misclassification percentage (%) curves. The nRMS error is computed from the RMS error between the targeted (t_i) and computed (c_i) values for i ranging from 1 to n, as described by Kumar et al. (2019a, 2019b):

$$RMS = \sqrt{\frac{1}{n} \sum_{i=1}^{n} (t_i - c_i)^2} \tag{6.5}$$

$$normalized\ RMS = \frac{RMS}{\sqrt{\frac{1}{n}}} \sum_{i=1}^{n} (t_i - mean)^2, \tag{6.6}$$

where the mean is given as:

$$mean = \frac{1}{n} \sum_{i=1}^{n} t_i. \tag{6.7}$$

The nRMS error curve demonstrates the overall error on the train and test sets, with a scale ranging from 0 to 1, where 0 corresponds to no error and 1 corresponds to highest error. It has been demonstrated that the lower the nRMS, the better the neural outcome (Kumar & Sain, 2018; Kumar et al., 2019a, 2019b, 2019c). The other benefit is that error performance can be observed from a single graph display.

The second performance evaluation parameter is the misclassification %. It is a quality control parameter to understand if incorrect predictions are made during the classification. To have control over this accuracy, it is recommended to create a truly representative distribution of observations for each class (i.e., target and non-target zones). The classification % is defined as the ratio of correct predictions to that of the total number of predictions.

$$Classification\ (\%) = \frac{Number\ of\ correct\ predictions}{Total\ numebr\ of\ predictions} x\ 100. \tag{6.8}$$

The misclassification percentage is defined as the ratio of wrong predictions to that of the total number of predictions.

$$Misclassification\ (\%) = \frac{Number\ of\ wrong\ predictions}{Total\ numebr\ of\ predictions} x\ 100. \tag{6.9}$$

To judge the performance evaluation of parameters of the intelligent neural model, we follow a general rule, i.e., the lower the nRMS, the lower the misclassification % (Kumar and Sain, 2018; Kumar et al., 2019a, 2019b, 2019c).

In the process of NN analysis, the only role of an interpreter is to provide correct example locations and logically confirm the quality of the intelligent model. The performance is further checked over a few key seismic lines. The model can then be ready to run over the entire data volume to fasten up the interpretation. This process of neural learning is called supervised neural learning where an engineered machine tries to follow teachings offered by the intelligent machine, i.e., the human brain, and generates an optimized output to the assigned task.

6.2.5. Validation

It is good practice to validate the results of a meta-attribute for strengthening the interpretation outcome. This validation can be done using borehole log signatures (if the prospect area is drilled with a well) or by comparing the outcome with available key published literature. For a chimney meta-attribute, which is designed and used for the interpretation of subsurface gas seepages, such inferences can be validated using certain key facts like: (a) density and neutron porosity logs; (b) seabed pock mark effects; (c) available reports on geo-microbial studies; and (d) past literatures.

6.3. RGB Blending and Geo-Body Extraction

Alves et al. (2015) demonstrated an extensive application of volume rendering (or geo-body extraction) and RGB blending to illuminate anomalous features from seismic volumes. The target areas are isolated by applying different opacity values to different amplitudes. Finally, the relevant geologic features are extracted in the form of a 3D object, called the geo-body, whose extraction is based on its opacity threshold value. Such a method can be applied for the identification of a geological feature, as long as its seismic character is distinct from adjacent (unrelated) strata. The success of these techniques largely depends on the selection of a range of seismic amplitudes that represent the target or object and an optimum color scale that embodies these ranges such that the surrounding country rock becomes transparent. The outputs of such technique are non-computer models and may not have any large significance towards qualitative and quantitative solutions of seismic interpretation problems. A common downside of this technique is the inability of an attribute (say seismic amplitude volume) that can hardly respond to a particular geologic target within the seismic volume. Barnes (2016) suggested that such interpretation practices could be improvised by using multi-attribute analysis approach or through some computer-oriented algorithms.

6.4. Summary

For the interpretation of seismic data, Kumar and Sain (2018) and Kumar et al. (2019a, 2019b, 2019c) demonstrated the merit of meta-attribute approach in lieu to geo-body extraction and RGB blending techniques. The meta-attributes are computer-generated models that are designed based on five basic steps: (i) seismic data conditioning; (ii) seismic attribute selection and extraction; (iii) example location selection; (iv) neural network operation; and (v) validation. The meta-attribute approach aims to deliver a realistic interpretation of target or object from the data. It must be admitted that the meta-attribute technique also possesses certain downsides when the data are noisy or are of poor quality. If such data, which cannot be improved by filtering, are taken as input for the meta-attribute computation, this may mislead the interpreter in identifying "target-yes or target-no" zones during training. This, in turn, may result in feeding inaccurate example locations to the network, which ultimately lowers the learning capability of the network and thus the efficiency of the NN fails. Of course, this is obvious for any approach.

Let us now look at some field examples in the upcoming chapters, where we will get an opportunity to learn more about how to compute the meta-attributes with a view to interpreting the geometry or architecture of subsurface features from 3D seismic data.

References

Al-Dossary, S., & Marfurt, K. J. (2007). Lineament-preserving filtering. *Geophysics, 72*(1), P1–P8. https://doi.org/10.1190/1.2387138

Alves, T. M., Omosanya, K. D., & Gowling, P. (2015). Volume rendering of enigmatic high-amplitude anomalies in southeast Brazil: A workflow to distinguish lithologic features from fluid accumulations. *Interpretation 3*: A1–A14. https://doi.org/10.1190/INT-2014-0106.1

Al-Shuhail, A. A., Al-Dossary, S. A., & Mousa, W. A. (2017). *Seismic data interpretation using digital image processing*. John Wiley & Sons. https://doi.org/10.1002/9781119125594

Aminzadeh, F., & De Groot, P. (2006). *Neural networks and other soft computing techniques with applications in the oil industry*. EAGE Publications.

Barnes, A. E. (2016). *Handbook of poststack seismic attributes*. Society of Exploration Geophysicists. https://doi.org/10.1190/1.9781560803324

Cartwright, J., & Møller Hansen, D. (2006). Magma transport through the crust via interconnected sill complexes. *Geology, 34*(11), 929–932. https://doi.org/10.1130/G22758A.1

Chopra, S., & Marfurt, K. J. (2007). Seismic attributes for prospect identification and reservoir characterization, SEG, Tulsa. https://doi.org/10.1190/1.9781560801900

Höcker, C., & Fehmers, G. (2002). Fast structural interpretation with structure-oriented filtering. *The Leading Edge, 21*(3), 238–243. https://doi.org/10.1190/1.1598121

Jaglan, H., Qayyum, F. & Huck, H. (2015). Unconventional seismic attributes for fracture characterization: *First Break, 33*, 101–109. https://doi.org/10.3997/1365-2397.33.3.79520

Kumar P. C., & Mandal, A. (2017). Enhancement of fault interpretation using multi-attribute analysis and artificial neural network (ANN) approach: A case study from Taranaki Basin, New Zealand. *Exploration Geophysics, 49*(3), 409–424. https://doi.org/10.1071/EG16072

Kumar, P. C., & Sain, K. (2018). Attribute amalgamation-aiding interpretation of faults from seismic data: An example from Waitara 3D prospect in Taranaki basin off New Zealand. *Journal of Applied Geophysics, 159*, 52–68. https://doi.org/10.1016/j.jappgeo.2018.07.023

Kumar, P. C., Omosanya, K. O., & Sain, K. (2019a). Sill Cube: An automated approach for the interpretation of magmatic sill complexes on seismic reflection data. *Marine and Petroleum Geology, 100*, 60–84. https://doi.org/10.1016/j.marpetgeo.2018.10.054

Kumar, P. C., Sain, K., & Mandal, A. (2019b). Delineation of a buried volcanic system in Kora prospect off New Zealand using artificial neural networks and its implications. *Journal of Applied Geophysics, 161*, 56–75. https://doi.org/10.1016/j.jappgeo.2018.12.008

Kumar, P. C., Omosanya, K. O., Alves, T., & Sain, K. (2019c). A neural network approach for elucidating fluid leakage along hard-linked normal faults. *Journal of Marine and Petroleum Geology, 110*, 518–538. https://doi.org/10.1016/j.marpetgeo.2019.07.042

Le Cun, Y. 1985. A learning procedure for asymmetric network. *Proc. Cognit. (Paris), 85*, 599–604.

Mukherjee, B., & Sain, K. (2019). Prediction of reservoir parameters in gas hydrate sediments using artificial intelligence (AI): A case study in the Krishna-Godavari basin (NGHP-02 Expedition), *Journal of Earth System Sciences, 128*, 199. https://doi.org/10.1007/s12040-019-1210-x

Mukherjee, B., & Sain, K. (2021). Vertical lithological proxy for gas hydrate sediments using statistical and artificial intelligence approach: A case study from Krishna-Godavari basin, offshore India (NGHP Expedition-02). *Marine Geophysical Research, 42*:3, 1–23.

Omosanya, K. O. (2018). Episodic fluid flow as a trigger for Miocene-Pliocene slope instability on the Utgard High, Norwegian Sea. *Basin Research. 1–23*. https://doi.org/10.1111/bre.12288

Parker, D. (1985). Learning-logic: Technical Report TR-47, Center for Computational Research in Economics and Management Science, MIT, April

Poulton, M. M. (Ed.) (2001). *Computational neural networks for geophysical data processing*. Elsevier, Amsterdam, The Netherlands.

Poulton, M. M. (2002). Neural networks as an intelligence amplification tool: A review of applications. *Geophysics, 67*(3), 979–993. https://doi.org/10.1190/1.1484539

Rosenblatt, F. (1962). Principles of neurodynamics. Perceptrons and the theory of brain mechanisms (No. VG-1196-G-8). Cornell Aeronautical Lab Inc Buffalo NY (1962).

Roberts, A. (2001). Curvature attributes and their application to 3-D interpreted horizons. *First Break, 19*, 85–100. https://doi.org/10.1046/j.0263-5046.2001.00142.x

Rumelhart, D. E., Durbin, R., Golden, R., & Chauvin, Y. (1995). Backpropagation: The basic theory. *Backpropagation: Theory, Architectures and Applications*, 1–34. https://doi.org/10.4324/9780203763247

Singh, A., Ojha, M., & Sain, K. (2020). Predicting lithology using neural network from ownhole data of a gas hydrate reservoir in Krishna-Godavari basin, eastern Indian offshore, *Geophysical Journal International, 220*(3), 1813–1837. https://doi.org/10.1093/gji/ggz522

Smallwood, J. R., & Maresh, J. (2002). The properties, morphology and distribution of igneous sills: modelling, borehole data and 3D seismic from the Faroe-Shetland area. *Geological Society London, Special Publication, 197*, 271–306. https://doi.org/10.1144/GSL.SP.2002.197.01.11

Tingdahl, K. M. (1999). Improving seismic detectability using intrinsic directionality. Earth Sciences Center, Goteborg University (Technical Report, B194). https://doi.org/10.1016/S0920-4105(01)00090-0

Tingdahl, K. M. (2003). Improving seismic chimney detection using directional attributes. In M. Nikravesh, F. Aminzadeh, L.A. Zadeh (Eds.), *Soft Computing and Intelligent Data Analysis in Oil Exploration, Developments in petroleum science* (Vol. 51, pp. 157–173). Elsevier. https://doi.org/10.1016/S0920-4105(01)00090-0

Tingdahl, K. M., & de Groot, P. F. (2003). Post-stack dip and azimuth processing. *Journal of Seismic Exploration, 12*, 113–126.

Tingdahl, K. M., & de Rooij, M. (2005). Semi-automatic detection of faults in 3D seismic data; *Geophysical Prospecting*, *53*, 533–542. https://doi.org/10.1111/j.1365-2478.2005.00489.x

Thomson, K., & Hutton, D. (2004). Geometry and growth of sill complexes: insights using 3D seismic from the North Rockall Trough. *Bulletin of Volcanology*, *66*, 364–375. https://doi.org/10.1007/s00445-003-0320-z

Werbos, P. (1974). Beyond regression: New tools for prediction and analysis in the behavioral sciences: PhD Dissertation, Applied Math, Harvard University, Cambridge, MA.

Part III
Case Studies Using Meta-Attributes

7

CHIMNEY INTERPRETATION

The delineation of gas migration pathways from the source rock through reservoirs to the seabed in the marine environment is important in unraveling potential over-pressured zones that are prone to drilling hazards. The migration pathways are associated with chaotic reflections, and help in understanding the petroleum system of a region. The chapter aims to illuminate these chaotic zones by designing a hybrid-attribute called the Chimney Cube meta-attribute using an artificial neural network. The case study is performed over the Maari prospect of Taranaki Basin, offshore New Zealand. The gas chimneys observed from the seismic data are analyzed using modern 3D visualization tools by displaying the chimney probability cube over different vertical seismic sections, horizon slices, and time slices, respectively.

7.1. Gas Chimneys: A Clue for Hydrocarbon Exploration

Gas chimneys are associated with chaotic reflections on seismic data, where reflectors are discontinuous and reflecting amplitudes are weaker. Delineation of chaotic vertical disturbances from data provides valuable information for hydrocarbon exploration and understanding petroleum system in a region. Further, the study can indicate potential over-pressured zones that help in avoiding drilling hazards (Heggland, 2004). However, the chaotic zones are often associated with noisy reflections that degrade the quality of seismic image (Heggland, 1997; Aminzadeh et al., 2002; Connolly & Garcia, 2012). Hence, there is a challenge for effective

Meta-Attributes and Artificial Networking: A New Tool for Seismic Interpretation,
Special Publications 76, First Edition. Kalachand Sain and Priyadarshi Chinmoy Kumar.
© 2022 American Geophysical Union. Published 2022 by John Wiley & Sons, Inc.
DOI: 10.1002/9781119481874.ch07

identification and mapping of disordered chaotic zones for fruitful subsurface imaging and interpretation towards the success of hydrocarbon exploration.

Over the past few decades, seismic attributes have been widely used for the delineation of subsurface geological features and hence interpretation (Al-Dossary & Marfurt, 2006; Bahorich & Farmer, 1995; Chen et al., 2008; Chopra & Marfurt, 2007; Farfour et al., 2015; Hardage et al., 1996a; Marfurt et al., 1998, 1999; Roberts, 2001; Tingdahl, 1999; Tingdahl & de Rooij, 2005; Tingdahl et al., 2001). Gas chimneys are generally associated with low similarity, low amplitude, and variable dip wipe-out zone, causing high-frequency attenuation due to scattering of seismic signals (Berndt et al., 2003; Brouwer et al., 2008, 2011; Connolly & Garcia, 2012; Ligtenberg, 2003; Petersen et al., 2010; Westbrook et al., 2008). The similarity, energy, dip variance, and frequency attributes are suitable for describing such chaotic and wipe-out zones from the surroundings (Brouwer et al., 2008; Connolly & Garcia, 2012). However, seismic interpreters encounter problems in extracting attributes and outlining the subsurface objects, as the same attributes may also correspond to several other geologic targets.

Meldahl et al. (1999) proposed a technique for improved interpretation of gas chimneys from seismic data. The approach is based on computation of multi-trace attributes and recombination of extracted attributes into one or two new attributes. This chapter demonstrates the interpretation of gas chimneys by computing a new attribute, called the Chimney Cube (CC) meta-attribute, by combining a set of seismic attributes based on the concept of ANN. To describe and demonstrate the methodology, the 3D seismic data in the Maari prospect off New Zealand has been utilized. The detailed description of this approach is available in the work by Singh et al. (2016).

7.2. Research Methodology

A workflow (Figure 7.1) has been designed to compute the CC meta-attribute from DSMF seismic data. In this steering, a statistical filter is applied to the seismic data volume (Figure 7.2a) that results in a smoothed seismic volume, known as the DSMF seismic volume (Figure 7.2b). This uses local dip and azimuth of seismic events for tracking locally with respect to the trace segments under investigation (Jaglan et al., 2015; Qayyum et al., 2015; Tingdahl, 1999; Tingdahl & de Groot, 2003; Tingdahl et al., 2001). The DSMF seismic volume is then used for extraction of attributes for further analysis (Figure 7.3).

Seismic attributes such as the similarity (Figure 7.4a), dip variance (Figure 7.4b), energy (Figure 7.4c), and frequency wash-out (Figure 7.4d) are selected for the delimitation of gas chimney. Other attributes such as the signal-to-noise ratio and two-way time are also taken into account. All these attributes are initially tested along a few key seismic lines to demonstrate their capability in capturing the

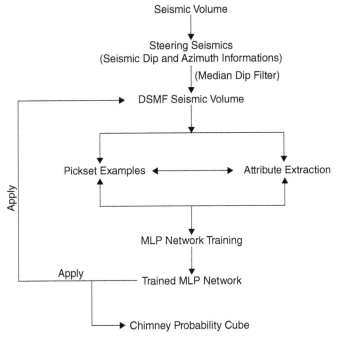

Seismic Volume

Steering Seismics
(Seismic Dip and Azimuth Informations)
(Median Dip Filter)

DSMF Seismic Volume

Apply

Pickset Examples ←→ Attribute Extraction

MLP Network Training

Apply   Trained MLP Network

Chimney Probability Cube

DSMF : Dip Steered Median Filter
MLP   : Multi-layered Perceptron

Figure 7.1 ANN-based workflow for generating Chimney Cube meta-attribute. Source: Singh, D., et. al., (2016). Interpretation of gas chimney from seismic data using artificial neural network: A study from Maari 3D prospect in the Taranaki basin, New Zealand. *Journal of Natural Gas Science and Engineering, 36, 339–357.*

responses from chimneys and demarcating them from the surroundings. Once the satisfactory outcome is achieved, the attributes set is chosen as an input to the ANN process.

Example locations are picked randomly from the data as "chimney-yes" and "chimney-no" zones (Figure 7.5) corresponding to binary values 1 and 0 to train the network. The chimney picks are taken at the most obvious locations of vertical hydrocarbon migration pathways, characterized by low-amplitude anomalies and chaotic zones. Non-chimney picks are taken at the locations without any migration pathways. Around 300 locations have been selected from the seismic volume for training.

The neural network learns from these examples and tries to establish a relationship between the input sets (seismic attributes) and the output (chimney-yes

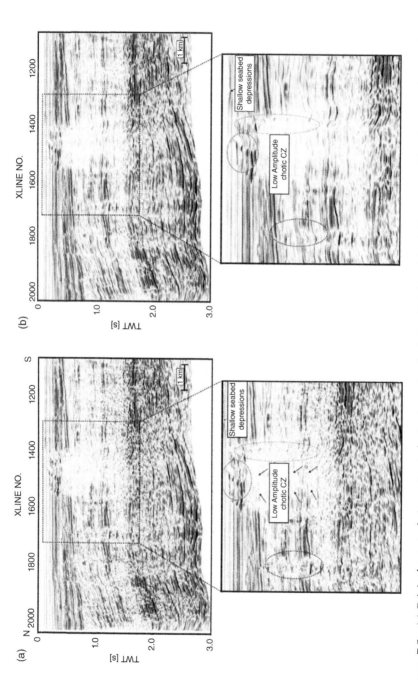

Figure 7.2 (a) Original pre-stack time-migrated seismic section for inline 693 from Maari 3D prospect. The chimneys (low amplitude, vertically aligned features), indicated by black arrows, are poorly visible. (b) Properly illuminated seismic image for the same section, conditioned using DSMF, shows signal enhancement at places where the images were poor (indicated by ovals and arrows in (a)) (CZ: Chimney Zone). Source: Singh, D., et. al., (2016). Interpretation of gas chimney from seismic data using artificial neural network: A study from Maari 3D prospect in the Taranaki basin, New Zealand. *Journal of Natural Gas Science and Engineering, 36*, 339–357.

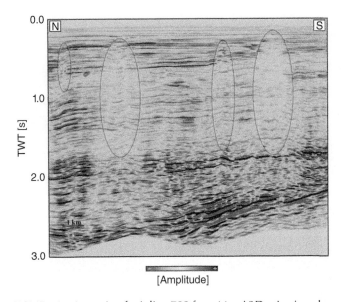

[Amplitude]

Figure 7.3 DSMF seismic section for inline 793 from Maari 3D seismic volume, depicting the chimney anomalies, marked by blue ovals. Source: Singh, D., et. al., (2016). Interpretation of gas chimney from seismic data using artificial neural network: A study from Maari 3D prospect in the Taranaki basin, New Zealand. *Journal of Natural Gas Science and Engineering, 36*, 339–357.

and chimney-no) (Figures 7.5 and 7.6). A fully connected multi-layer perceptron (MLP) is used to perform the neural training. The training is carried out iteratively by adjusting the network parameters (learning rate, momentum, and most importantly the weights of the connections) unless a minimum error and minimum misclassification % between the prediction and identified response is achieved. The training results are visualized on key seismic lines for quality check (QC). Once satisfied with the QC, the neural network is run over the entire seismic volume to obtain a probability cube called the chimney probability volume or simply the chimney cube meta-attribute. The meta-attribute contains values between 0 and 1, where 0 corresponds to the lowest probability of chimneys and 1 corresponds to the highest probability of chimneys. The chimney volume is viewed by overlying on seismic sections and time slices.

7.3. Chimney Validation

The presence of chimney can be validated in three different ways as described in the following sections.

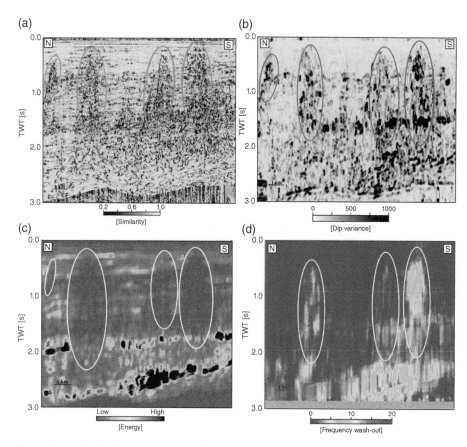

Figure 7.4   Vertical seismic section along inline 793 from Maari 3D prospect for (a) Dip-steered similarity attribute, portraying chimneys associated with low similarity values, marked by orange ovals; (b) variance dip attribute, showing chimney zones (red ovals), associated with high variance dips; (c) energy attribute, displaying the chimney zones (yellow ovals), associated with low energy; (d) frequency wash-out attribute for inline, depicting chaotic chimney zones (yellow ovals), associated with high-frequency attenuation. Source: Singh, D., et. al., (2016). Interpretation of gas chimney from seismic data using artificial neural network: A study from Maari 3D prospect in the Taranaki basin, New Zealand. *Journal of Natural Gas Science and Engineering, 36*, 339–357.

7.3.1. Geological Validation

Geological knowledge from available information is important for validation. The geological tops, as observed from the drilled well(s), are projected over the seismic volume. The chimney meta-attribute, co-rendered with seismic attribute(s), is then examined for plausible geological features such as faults and their intersections over these formations with a view to ascertaining the presence of gas seepage and accumulation.

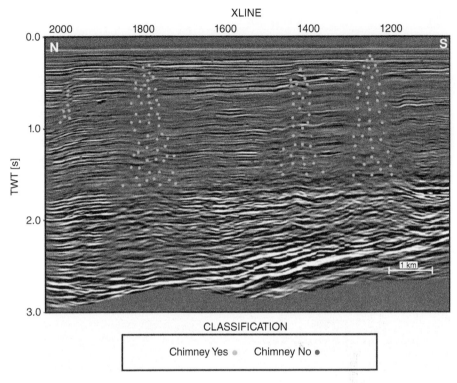

Figure 7.5 Representative seismic section showing the example locations of pickset along inline 793, which are classified into chimney-yes (green dots) and chimney-no (blue dots) groups to train the MLP network. Source: Singh, D., et. al., (2016). Interpretation of gas chimney from seismic data using artificial neural network: A study from Maari 3D prospect in the Taranaki basin, New Zealand. *Journal of Natural Gas Science and Engineering, 36,* 339–357.

7.3.2. Petrophysical Validation

The petrophysical properties from different well logs, e.g., the density (RHOB) and neutron porosity (NPHI), can be assessed for gas-related effects for correlating chimney meta-attribute with the petrophysical logs from the drilled Moki-1 well (TEOL, 1984).

7.3.3. Soft Sediment Deformation Anomalies

The effects of gas seepage, venting of oil, or mobilized sediments through the seabed are observed in the form of markers such as the pockmarks, mud diapirs, and mud volcanoes (Løseth et al., 2009). Observation of these features over the seismic data (sections, time slices, horizon slices) validates the presence of chimneys.

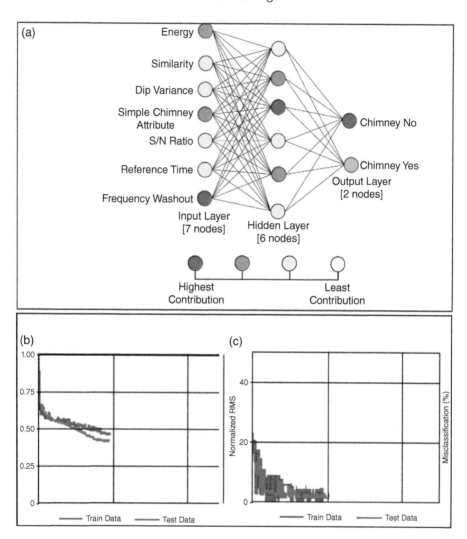

Figure 7.6 (a) The MLP network consisting of input layer, hidden layer, and output layer. Each layer consists of different nodes interconnected with each other; (b) nRMS error between the observed and predicted outcome for the train (red) and test (blue) data sets; (c) Misclassification % for the train (red) and test (blue) data sets, respectively.

7.4. Interpretation Using Chimney Cube

The MLP network (Figure 7.6a) consists of 15 fully connected nodes; 7, 6, and 2 nodes are associated with the input, hidden, and output layers, respectively. The activation function used here is a sigmoid function that is continuous, monotonically increasing, differentiable, and bounded. It takes the input and squashes the output in terms of 0s and 1s, where 0 refers to "chimney-no" and 1 refers to "chimney-yes." The network performs its operation by randomly splitting the picked data into 70% for training and remaining 30% for testing. We observe that after 10 iterations, the nRMS error for both the train and test data sets attain the minimum value of 0.4 and 0.6, respectively (Figure 7.6b). Corresponding to this, a minimum misclassification (%) of 5.08 and 10.26 % is also obtained for the train and test data sets, respectively (Figure 7.6c).

The chimney probability volume obtained from this neural training by the computation of chimney cube meta-attribute is displayed on vertical cross seismic sections (Figure 7.7). The chimney output in vertical seismic section exhibits the extent of active gas leakage from the source through the reservoirs. This observation is more prominent when the chimney output is overlain on the seismic cross-section (Figure 7.7b).

The chimneys are originated from the source rocks of the Pakawau Group belonging to the Late Cretaceous period, charging into the Kapuni Group of the Paleocene to the Eocene period and propagating through the Miocene reservoirs all the way to the seabed. To see the gas migration we have created several time slice or horizons. Co-rendering the chimney with similarity attributes at 2,500 ms time slice (Figure 7.8a) cuts the Pakawau Group, and shows the presence of gas accumulation. High probable chimneys are represented by the deep yellow color and low probable chimneys are indicated by the green to blue color. This formation is also associated with several polygonal fault systems that are observed close to the gas zones. Development of layer bound fault systems within this formation provides a pathway for the gas to be charged into the overlying Kapuni Group. Co-rendering the chimney and similarity attributes at the 1,660 ms time slice (Figure 7.8b) that cuts the Kapuni Group also reveals the fault system and presence of gas accumulation.

Most of the gas deposits are concentrated along the faults, as are observed on the horizon slices over the top of Kapuni Formation (Figure 7.9a). The horizon slice prepared by co-rendering the chimney and similarity attributes over the top of Mohakatino Formation (Miocene reservoir) shows that gas chimneys are distributed in patches and are more pronounced in the NE and SE parts of the region (Figure 7.9b). The Mohakatino Formation is deformed and highly faulted, characterized by low similarity. However, zones within these faults are characterized by high chimney values. The fault zones in the extreme northern, NW, and SE parts of the formation are associated with high chimney probability. Fault intersections are observed in the NW, and the zones within these intersections are characterized by high chimney values. These observations infer not only the faulting system of the formation but also imply that these faults act as weak zones or conducive paths for gas seepage. The time slice (Figure 7.10a) at 560 ms that cuts the Mohakatino Formation also honors these observations. Some of the faults in the extreme northern and NW part over this time slice intersect with

(a)

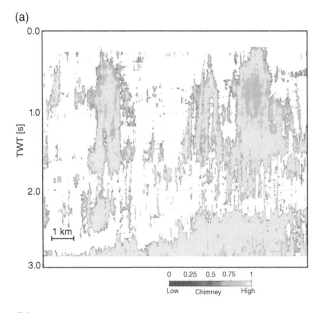

(b)

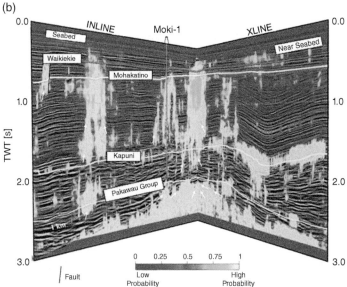

Figure 7.7 (a) The chimney meta-attribute for inline 793. High chimney probabilities are indicated by deep yellow color, whereas low probabilities are indicated by blue color. (b) Chimney attributes in the vertical section for inline 793 and crossline 1253, showing the gas chimneys moving upwards from the source rock through the reservoirs (white arrows) to the seabed (red arrows). Location of Moki-1 well (time-converted) is also shown over the seismic section in (b). Source: Singh, D., et. al., (2016). Interpretation of gas chimney from seismic data using artificial neural network: A study from Maari 3D prospect in the Taranaki basin, New Zealand. *Journal of Natural Gas Science and Engineering, 36,* 339–357.

(a)

(b)

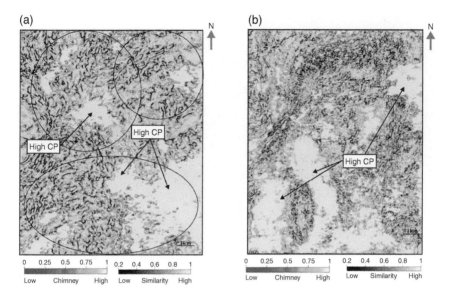

Figure 7.8 (a) Chimney and similarity attributes, co-rendered at 2500 ms time slice that cuts the Pakawau Group, show the polygonal fault system throughout the formation, which are associated with high gas chimneys (black ovals). (b) Chimney and similarity attributes, co-rendered at 1660 ms time slice that cuts the Kapuni Group, also show the faults and gas accumulation (CP: Chimney Probability). Source: Singh, D., et. al., (2016). Interpretation of gas chimney from seismic data using artificial neural network: A study from Maari 3D prospect in the Taranaki basin, New Zealand. *Journal of Natural Gas Science and Engineering*, *36*, 339–357.

each other, and higher chimneys are observed at these zones. Chimney patches with higher probability are observed over the NE and SE parts of the formation. Similar analysis is also carried over the time slice (Figure 7.10b) at 440 ms that cuts the Waikiekie Formation, which exhibits patchy distribution of chimneys in the SW, SE, and NE parts of the formation. Sharp signatures of faults, characterized by low similarity, are observed over this formation. The faults are also associated with high chimney patches. However, the faults observed in the NW part of the formation do not show signatures of gas chimneys, indicating that the faults are sealed in the NW part.

The time slice at 152 ms over the seabed, co-rendered with the chimney and similarity attributes, shows the signature of gas migration (Figure 7.11a). The pockmarks (circular appearance), observed over the seabed, are associated with low similarity and high chimney attributes. These features are nothing but the seafloor depressions caused by the escape of fluids and gases through them (Petersen et al., 2010). To see the signature at depth, the seismic section along the random line AB passing through these circular features is taken, and the time

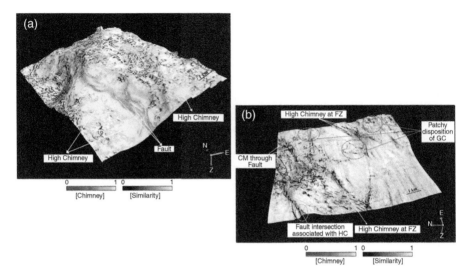

Figure 7.9   (a) Chimney and similarity attributes, co-rendered at horizon slice over the top of Kapuni Formation. Chimney effects (high probability) are observed along the fault zones (low similarity). High faulting activities along with several weak zones are observed within the formation. The weak zones act as conducive pathways for gas seepage through them. (b) Chimney and similarity attributes, co-rendered at horizon slice over the top of Mohakatino Formation. Patchy distributions of gas chimneys are mostly observed towards the NE and SE parts. Fault intersection zones accompanied by high chimney are highlighted (orange oval) (FZ: Fault Zone; CM: Chimney Migration; GC: Gas Chimney; HC: High Chimney). Source: Singh, D., et. al., (2016). Interpretation of gas chimney from seismic data using artificial neural network: A study from Maari 3D prospect in the Taranaki basin, New Zealand. *Journal of Natural Gas Science and Engineering, 36,* 339–357.

section (Figure 7.11b) also shows the gas escape features from the shallow reservoirs all the way to the seabed. Since the density and neutron porosity logs are used for petrophysical correlation, the location of Moki-1 well is shown over the seismic section in Figures 7.7b, where gases are encountered at depths ranging from 1,100 to 2,600 (~1.15–2.25 s TWT) meters below kelly bush (mbkb) (Figure 7.12). These are also indicated by a decrease in observed density and porosity logs. The well was drilled up to a depth of 2,620 meter below Kelly Bush (mbkb). No information is available below this depth as well as above 650 mbkb. Thus, the findings based on seismic attribute studies are corroborated very well by the drilling results.

The chimney outcome from the entire study area has been summarized in Figure 7.13, which shows that the gas clouds originating from the Pakawau Group (source rock) of the Late Cretaceous age propagate through the overlying Kapuni

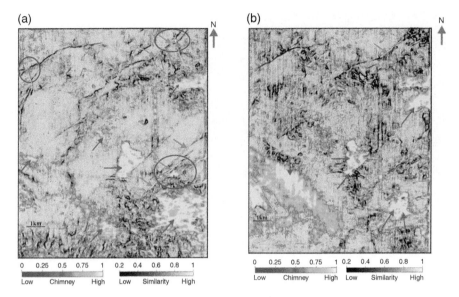

Figure 7.10 (a) Chimney and similarity attributes, co-rendered at 560 ms time slice that cuts the Mohakatino Formation. Chimneys are distributed in patches (purple arrows). Blue ovals in the NE, NW, and eastern parts, highlighting fault intersection (violet arrows), are characterized by high chimney effects. (b) Chimney and similarity attributes, co-rendered at 440 ms time slice that cuts the Waikiekie Formation. Gas chimneys (violet arrows) are observed in the central and NE parts of the formation. Source: Singh, D., et. al., (2016). Interpretation of gas chimney from seismic data using artificial neural network: A study from Maari 3D prospect in the Taranaki basin, New Zealand. *Journal of Natural Gas Science and Engineering, 36*, 339–357.

Group of the Eocene and Wai-iti Group (Mohakatino and Waikiekie Formations) of the Miocene reservoirs to the seabed. Frequency attenuation and signal deterioration between the deep and shallow reservoirs reveal that gas chimneys extend through the formation to the seabed, which is indicated by pockmarks or wipe-out zones. Thus, the workflow, described in this chapter, helps in understanding the petroleum system in the study area, and provides a measure for mitigating hazards caused by overpressure due to the presence of gas.

The MLP network, which is a supervised learning process, learns from example sets (i.e. chimney sets in this case) to classify chimney-yes and chimney-no zones from seismic data. The advantages of this approach is that it uses the interpreter's ability, insight, and knowledge while training the system from the example sets over a small volume of data, and then automatically delineates the feature (here gas chimneys or migration pathways) of interest from the whole volume of data with limited intervention by the human analysts.

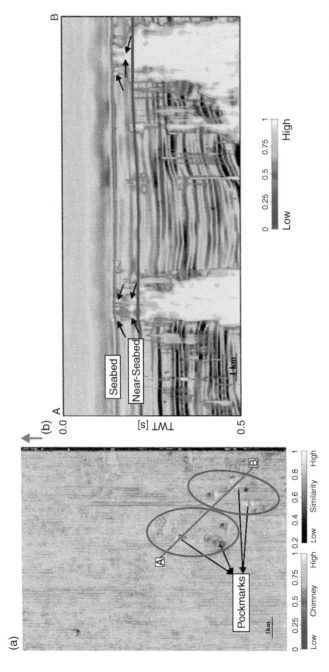

Figure 7.11 (a) Chimney and similarity attributes, co-rendered at 152 ms time slice for the seabed, exhibiting pockmarks (circular morphology), indicated by blue dotted ovals that are associated with high chimney probability and gas seepage to the seabed. (b) Seismic section up to 0.5 s along the random line AB passing through the circular features as shown in (a), exhibiting the chimney attribute and vertical gas migrations (black arrows) from the source to the seabed through the near-seabed. Source: Singh, D., et. al., (2016). Interpretation of gas chimney from seismic data using artificial neural network: A study from Maari 3D prospect in the Taranaki basin, New Zealand. *Journal of Natural Gas Science and Engineering, 36*, 339–357.

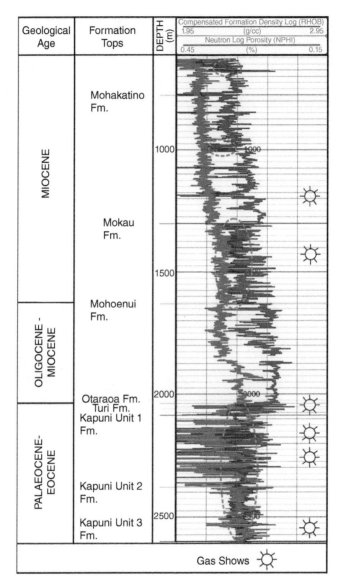

Figure 7.12   The density and neutron porosity logs at Well Moki-1 passing through the study area. The decreasing trend of both logs (green depth interval) in which gas shows were encountered are marked by circular symbols.

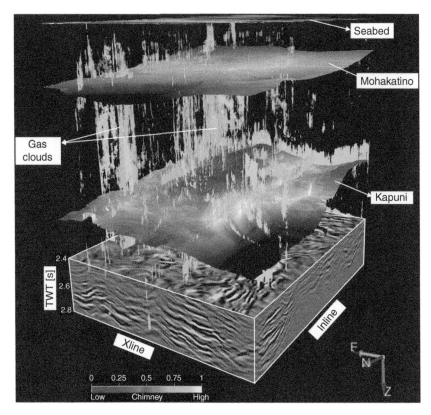

Figure 7.13   3D volumetric visualization of gas clouds or chimneys showing its rise from the thermally mature source rock, propagating through the Eocene (Kapuni Formation) and Miocene (Mohakatino Formation) reservoirs all the way to the seabed. Source: Singh, D., et. al., (2016). Interpretation of gas chimney from seismic data using artificial neural network: A study from Maari 3D prospect in the Taranaki basin, New Zealand. *Journal of Natural Gas Science and Engineering, 36, 339–357.*

7.5. Summary

The following are the major conclusions drawn from this case study:
- A workflow based on a neural network is proposed for the computation of a hybrid attribute called the chimney cube (CC) meta-attribute from a suit of other seismic attributes.
- The CC meta-attribute has been very efficient in discriminating gas chimneys from the surroundings.

- Application of this approach to the time-migrated 3D seismic data in the highly structured and deformed Maari field of the Taranaki Basin off New Zealand has delimited the gas chimney or gas clouds very well.
- The study has demonstrated that gas has originated from the Late Cretaceous source rocks and seeped through the overlying Eocene to Miocene to Pliocene to recent formations, the imprints of which are observed as pockmarks on the seabed.
- Results have been validated with the available well log information.

References

Al-Dossary, S., & Marfurt, K. J. (2006). 3-D volumetric multi-spectral estimates of reflector curvature and rotation. *Geophysics*, *71*, 41–51. https://doi.org/10.1190/1.2242449

Aminzadeh, F., Connolly, D., Heggland, R., & de Groot, P.F.M (2002). Geo-hazard detection and other application of chimney cubes. *The Leading Edge*, *21*, 681–685. https://doi.org/10.1190/1.1497324

Bahorich, M., & Farmer, S. (1995). 3-D seismic discontinuity for faults and stratigraphic features: The coherence cube. *The Leading Edge*, *14*, 1053–1058. https://doi.org/10.1190/1.1437077

Berndt, C., Bunz, S., & Mienert, J. (2003). Polygonal fault systems on the Mid-Norwegian margin: a long-term source for fluid flow. In Rensbergen, P.V., Hills, R., Maltman, A., Morley, C. (Eds.), *Origin, processes, and effects of subsurface sediment mobilization on reservoir to regional scale*. Geological Society of London, Special Publication. https://doi.org/10.1144/GSL.SP.2003.216.01.18

Brouwer, F. G. C., Welsh, A., Connolly, D. L., Selva, C., Curia, D., & Huck, A. (2008). *High Frequencies Attenuation and Low Frequency Shadows in Seismic Data Caused by Gas Chimneys, Onshore Ecuador*. Paper presented in 70th EAGE Conference and Exhibition incorporating SPE EUROPEC 2008. https://doi.org/10.3997/2214-4609.20147600

Brouwer, F., Tingdahl, K., & Connolly, D. (2011). *A Guide to the Practical Use of Neural Networks*. Paper presented in 31st Annual Gulf Coast Section SEPM Foundation Bob F. Perkins Research Conference, Houston, Texas, 4–7 Dec. 2011. https://doi.org/10.5724/gcs.11.31.0440

Chen, G., Matteucci, G., Fahmi, B., & Finn, Ch. (2008). Spectral-decomposition response to reservoir fluids from a deepwater West Africa reservoir. *Geophysics*, *73*, 23–30. https://doi.org/10.1190/1.2978337

Chopra, S., & Marfurt, K. J. (2007). Seismic attributes for prospect identification and reservoir characterization. SEG, Tulsa. https://doi.org/10.1190/1.9781560801900

Connolly, D., & Garcia, R. (2012). GEOLOGY & GEOPHYSICS-Tracking hydrocarbon seepage in Argentina's Neuquén basin. World Oil, 101–104.

Farfour, M., Yoon, W. J., & Kim, J. (2015). Seismic attributes and acoustic impedance inversion in interpretation of complex hydrocarbon reservoirs. *Journal of Applied Geophysics*, *114*, 68–80. https://doi.org/10.1016/j.jappgeo.2015.01.008

Hardage, B. A., Carr, D. L., Lancaster, D. E., Simons, J. L., Hamilton, D. S., Elphick, R. Y., Oliver, K. L., & Johns, R. A. (1996a). 3D seismic imaging and seismic attribute analysis of genetic sequences situated in low accommodation conditions. *Geophysics*, *61*, 1351–1362. https://doi.org/10.1190/1.1444058

Heggland, R. (1997). Detection of gas migration from a deep source by the use of exploration 3D seismic data. *Marine Geology*, *137*, 41–47. https://doi.org/10.1016/S0025-3227(96)00077-1

Heggland, R. (2004). Definition of geohazards in exploration 3-D seismic data using attributes and neural-network analysis. *AAPG Bulletin*, *88*, 857–868. https://doi.org/10.1306/02042004

Jaglan, H., Qayyum, F., & Huck, H. (2015). Unconventional seismic attributes for fracture characterization: *First Break*, *33*, 101–109. https://doi.org/10.3997/1365-2397.33.3.79520)

Ligtenberg, J. H. (2003). Unravelling the petroleum system by enhancing fluid migration paths in seismic data using a neural network based pattern recognition technique. *Geofluids*, *4*, 255–261. https://doi.org/10.1046/j.1468-8123.2003.00072.x

Løseth, H., Gading, M., & Wensaas, L. (2009). Hydrocarbon leakage interpreted on seismic data. *Marine and Petroleum Geology*, *26*(7), 1304–1319. https://doi.org/10.1016/j.marpetgeo.2008.09.008

Marfurt, K. J., Kirlin, R.L., Farmer, S.L. & Bahorich, M.S. (1998). 3-D seismic attributes using a semblance-based coherency algorithm. *Geophysics*, *63*, 1150–1165. https://doi.org/10.1190/1.1444415

Marfurt, K. J., Sudhaker, V., Gersztenkorn, A., Crawford, K. D., & Nissen, S. E. (1999). Coherency calculations in the presence of structural dip. *Geophysics*, *64*, 104–111. https://doi.org/10.1190/1.1444508

Meldahl, P., Heggland, R., Bril, B., & de Groot, P. (1999). *The chimney cube, an example of semi-automated detection of seismic objects by directive attributes and neural networks: Part I;* methodology. Paper presented in SEG Annual Meeting, Houston, Expanded Abstracts, 1, 931–934. https://doi.org/10.1190/1.1821262

Petersen, C. J., Bunz, S., Hustoft, S., Mienert, J., & Klaeschen, D. (2010). High-resolution P-Cable 3D seismic imaging of gas chimney structures in gas hydrate sediments of an Arctic sediment drift. *Marine and Petroleum Geology*, *27*, 1981–1994. https://doi.org/10.1016/j.marpetgeo.2010.06.006

Qayyum, F., Catuneanu, O., & Bouanga, C. E. (2015). Sequence stratigraphy of a mixed siliciclastic-carbonate setting, Scotian Shelf, Canada: *Interpretation*, SEG, *3*, pp. SN21–SN37. https://doi.org/10.1190/INT-2014-0129.1

Roberts, A. (2001). Curvature attributes and their application to 3 D interpreted horizons. *First Break*, *19*, 85–100. https://doi.org/10.1046/j.0263-5046.2001.00142.x

Singh, D., Kumar, P. C., & Sain, K. (2016). Interpretation of gas chimney from seismic data using artificial neural network: A study from Maari 3D prospect in the Taranaki basin, New Zealand. *Journal of Natural Gas Science and Engineering*, *36*, 339–357. https://doi.org/10.1016/j.jngse.2016.10.039

TEOL (Tricentrol Exploration Overseas Ltd.) 1984. Well Completion Report Moki-1 PPL 38114. Ministry of Economic Development New Zealand. Unpublished Petroleum Report Series, PR987.

Tingdahl, K. M. (1999). Improving seismic detectability using intrinsic directionality; Technical Report, Earth Sciences Center, Goteborg University, B194.

Tingdahl, K. M., Bril, A. H., & de Groot, P. (2001). Improving seismic chimney detection using directional attributes. *Journal of Petroleum Science and Engineering, 29*, 205–211. https://doi.org/10.1016/S0920-4105(01)00090-0

Tingdahl, K. M., & de Groot, P. E. M. (2003). Post-stack dip and azimuth processing; *Journal of Seismic Exploration, 12*, 113–126.

Tingdahl, K. M., & de Rooij, M. (2005). Semi-automatic detection of faults in 3D seismic data; *Geophysical Prospecting, 53*, 533–542. https://doi.org/10.1111/j.1365-2478.2005.00489.x

Westbrook, G. K., Exley, R., Minshull, T., Nouze, H., Gailler, A., Jose, T., Ker, S., & Plaza, A., (2008). High-resolution 3D Seismic Investigations of Hydrate-bearing Fluid-escape Chimneys in the Nyegga Region of the Vøring Plateau, Norway. In *Proceedings of the 6th International Conference on Gas Hydrates*, Vancouver, British Columbia, July 6–10, pp. 8.

8

FAULT INTERPRETATION

Several structural attributes have evolved over time and made progressive improvement in the delineation of faults and discontinuous structures that have made interpretation easier. The chapter aims to showcase the approach of designing an optimized Thinned Fault Cube meta-attribute by fusing a number of appropriate attributes through an artificial neural network. The case study is performed over the Waitara 3D prospect of Taranaki basin, offshore New Zealand, and it demonstrates a robust and automatic interpretation of subsurface faults.

8.1. Fault Meta-Attribute: A Motivation

Seismic attributes have revolutionized the interpretation of faults from seismic data. Different sets of structural attributes such as the coherency and semblance attributes (Bahorich & Farmer, 1995; Gresztenkorn & Marfurt, 1999; Marfurt et al., 1998), curvature attribute (Al-Dossary & Marfurt, 2006; Chopra and Marfurt, 2007; Roberts, 2001), and dip-azimuth attribute (Barnes, 2003; Chopra & Marfurt, 2007; Dalley, 2008; Luo et al., 2002) have evolved over time and progressively improved the images of subsurface faults and related structures that have made interpretation easier.

The high-performance computational facilities could enable the generation of structural attributes like similarity, dip angle variance, data adaptive azimuth ridge enhancement filter (REF), fault likelihood, flexures, etc. (Brouwer & Huck, 2011; de Groot, 1995, 1999; de Rooij & Tingdahl, 2002; Di & Gao, 2016; Hale, 2013; Tingdahl, 2003; Tingdahl & de Groot, 2003; Tingdahl & de Rooij, 2005; Wu, 2017; Wu

Meta-Attributes and Artificial Networking: A New Tool for Seismic Interpretation,
Special Publications 76, First Edition. Kalachand Sain and Priyadarshi Chinmoy Kumar.
© 2022 American Geophysical Union. Published 2022 by John Wiley & Sons, Inc.
DOI: 10.1002/9781119481874.ch08

and Hale, 2016;). However, there is no unique attribute that can lead to unequivocal interpretation of a particular geologic feature such as the fault. Thus, there is a need to look for a single attribute, which can delineate the subsurface geologic body or physical property with great certainty. This has motivated the search for a hybrid- or meta-attribute by amalgamating several other attributes that can suitably capture the target and ease the process of interpretation from a large volume of data.

To demonstrate this concept, 3D time-migrated seismic data from the Waitara prospect in the Taranaki Basin off New Zealand have been used. The high-resolution data volume consisting of 503 inlines (N-S) and 3,116 xlines (E-W), covering an area of 120 km^2, was acquired with bin size of 25.0 m × 12.5 m at vertical sampling rate of 2 ms (equivalent to Nyquist frequency of 250Hz) and 8 s record length. The key objectives behind the seismic data acquisition were to image the primary targets of the Eocene and Miocene Formations, deeper faulted Cretaceous formations and the signatures of mushroom-shaped volcanic bodies (Todd Energy, 2014). The acquired data has been processed by conventional processing flows, e.g., trace edit, geometry update, deconvolution, linear noise attenuation, anomalous amplitude attenuation (AAA), and velocity analysis followed by migration. The seismic volume is displayed in the Society of Exploration Geophysicists (SEG) normal polarity convention. Hence, a trough (red reflection) represents a downward decrease in acoustic impedance and a peak (black reflection) signifies an increase in acoustic impedance.

In this study, three sets of seismic attributes have been computed from seismic data and united into a hybrid attribute based on the ANN approach to derive three optimized Fault Cube (FC) meta-attributes for capturing the subsurface faults. The outcomes of three different cases are compared with each other to predict an optimal case for the best possible design of workflow to improve the interpretation of subsurface faults.

8.2. Research Methodology

A workflow that explains the data conditioning, which is the foremost requirement for extraction of seismic attributes in designing the meta-attribute, is shown in Figure 8.1. Seismic attributes have been categorized into three different cases (Case-I; Case-II; Case-III) based on their evolution, computational efficacy, and ability to capture seismic responses from geologic faults and discontinuities.

The data is structurally conditioned using a pre-processed steering cube that stores the dip-azimuth information at each sample location. The structure-oriented filters (SOF) are employed to improve the structural prominence of discontinuous structures and to reduce the unwanted parts from the data, i.e. the noise. The filter operates on (i) orientation of seismic reflectors, (ii) identification of reflection terminations, and (iii) preservation of these terminations for smoothing the

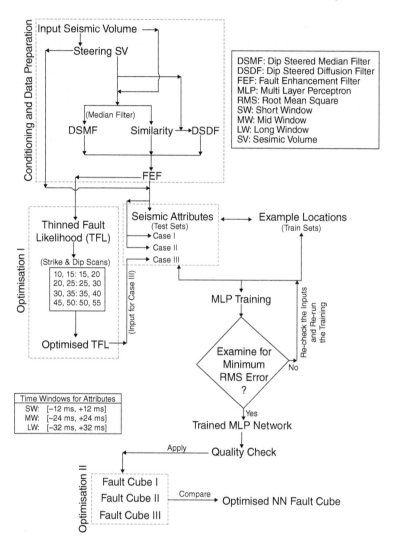

Figure 8.1 Workflow with different parts such as data conditioning, attribute selection and extraction, and neural operations.

reflectors. The seismic volume is filtered initially to obtain the DSMF and DSDF seismic volumes, and both are logically merged using a cut-off value of 0.5 with the help of the similarity attribute. This generates a Fault-Enhanced Filter (FEF) that performs edge preservation and sharpening operations. This data volume along with pre-processed steering data is used as inputs for attribute extraction and performing further analysis.

Different structural, physical, and geometrical seismic attributes are segregated in three different cases (Cases I, II, and III). Once the attributes are selected, it is crucial to parametrize them in such a way that can offer efficient delineation of geologic target from the data. Care is taken to assign the inline and xline step-outs and time gates (or window lengths). The coherency, semblance, and similarity attributes are extracted using three separate time window lengths: long (64ms), mid (48ms), and short (24ms). The window lengths are so designed to match the orientation and vertical extent of target (e.g., faults and discontinuities) throughout the data volume, and help in improved identification of seismic attributes. The thin fault likelihood (TFL) attribute is extracted from the FEF seismic scanned for a range of fault strikes and dips to recognize the maximum likelihood of faults and associated discontinuities. The likelihood attribute is computed with 10 to 50 scans (at every 5 incremental step) in the strike direction within a range of 0 to 360 degrees and 15 to 55 scans (at every 5 incremental steps) in the dip direction within a range of 35 to 85 degrees. A wider range of dip and strike scans is performed to apprehend the maximum structural details from the data volume. The extracted attributes are quality checked to test their efficacy in capturing the geologic target. The attributes are then computed over the entire seismic volume to prepare the volumetric seismic attributes that act as inputs for neural computation.

Example locations are prepared to feed the neural system for all three cases such that the network receives guidance within the input attribute sets for discriminating the characters of fault and non-fault zones. The fault picks are made at zones associated with the bed displacements and reflector breaks, and are characterized by low similarity, low coherency, and variable dip anomaly attributes. Non-fault picks are made at zones that are devoid of such disturbances. Around 2,850 example locations (1,400 fault picks and 1,450 non-fault picks for every 10th seismic line) are selected from the seismic data volume.

Neural training is performed to establish a relationship between the input (seismic attributes) and output (fault-yes and fault-no) sets. The training is performed through a fully connected MLP network that consists of 14 (Case-I), 16 (Case-II), and 15 (Case-III) fully connected nodes, respectively. The activation function used here is a sigmoid function that squashes the output in terms of 0s and 1s, where 0 refers to "fault-no" and 1 refers to "fault-yes." The network performs its operation by randomly splitting a small volume (15-20%) of total data, and "fault-yes" and "fault-no" are assigned to each node. 70% of chosen data are used for training and the remaining 30% are used for testing. The training is carried out iteratively to minimize the nRMS error and misclassification % between the predicted and identified outcome by adjusting or updating the rate of learning, momentum, and most importantly the weights (Atakulreka & Sutivong, 2007; Rosenblatt, 1962; Rumelhart et al., 1995; Singh et al., 2016). Once the error is minimized, the result is quality checked over a few randomly selected seismic lines. The NN is then applied over the entire seismic volume to obtain the probability

FC meta-attribute. The results of fault attributes are then validated with the available regional geology or well data or other information. The result of the probability cube is also compared with that of an individual attribute to infer the superiority of the meta-attribute over an individual attribute in delineating the geologic target of interest. The FC volume contains the samples whose values range from 0 to 1 (statistical probabilities), where 0 signifies the zero probability of faults and 1 signifies the maximum probability of being a fault. The probability FC is visualized over seismic sections and time slices.

8.3. Results and Interpretation

Structural conditioning of original time-migrated seismic data (Figure 8.2a) shows enhanced visibility of structural discontinuities or faults (Figure 8.2b). The DSMF seismic smoothes the reflectors by eliminating the noise and enhancing the continuity of events. The FEF seismic data (Figure 8.2c) has sharpened the structures by refining the signal strength near the fault zones. The conditioned data is then used for extraction of seismic attributes and better illumination of subsurface.

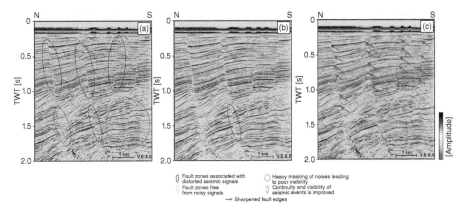

Figure 8.2 (a) Original time-migrated section for inline 1078 from 3D seismic volume with disturbed seismic signals (black and red dotted ovals). Noisy reflections have obscured the fault. (b) The DSMF conditioned section demonstrating improved seismic signal with fault zones free from random noises (pink and green dotted ovals). (c) The FEF seismic section illustrating improved geological features. Reflector edges within the fault zones are sharpened and distinct (red arrows). Source: Kumar, P. C., & Sain, K. (2018). Attribute amalgamation-aiding interpretation of faults from seismic data: An example from Waitara 3D prospect in Taranaki basin off New Zealand. *Journal of Applied Geophysics, 159*, 52–68.

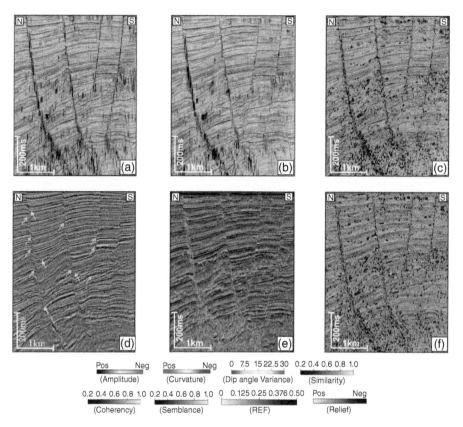

Figure 8.3 Co-rendered seismic sections for inline 1078, displaying amplitude data with extracted attributes: (a) coherency, (b) semblance, (c) similarity, (d) curvature, (e) dip variance attribute, and (f) REF attributes. Source: Kumar, P. C., & Sain, K. (2018). Attribute amalgamation-aiding interpretation of faults from seismic data: An example from Waitara 3D prospect in Taranaki basin off New Zealand. *Journal of Applied Geophysics, 159*, 52–68.

The low values of coherency and semblance attributes (Figures 8.3a, b) characterize the faults. The similarity attribute (Figure 8.3c) shows improved visibility of discontinuities compared to the coherency and semblance attributes (Figures 8.3a, b). The up-thrown portion of the faults, anticlinal (associated with positive values), and synclinal or bowl-shaped (associated with negative values) features are prominently observed by the curvature attribute (Figure 8.3d). Moreover, the attributes show minor bed displacements and systems of discontinuous structures in the deeper portion (~1.7–2.2 s TWT). Structural discontinuities are also associated with variable dip that changes from high to low along and either side of the fault zones, as illustrated by the dip variance attribute in Figure 8.3e. The REF attribute (Figure 8.3f) also shows distinct images of discontinuous features from the shallower to the deeper part, which appears smeared in Figures 8.3a, b.

The optimized TFL attribute (Hale, 2013; Wu & Hale, 2016) captures the structural details when scanned for maximum likelihood with several steps in the strike and dip directions (Figure 8.4a–i). It is observed that by increasing the scans from step 10/15 to 50/55 in the respective strike and dip direction, the geologic faults and minute structural details are apprehended by the TFL attribute (Figure 8.4i). These observations account for an optimized case of the TFL attribute (i.e. TFL with number of strike and dip scans: 50,55) that demonstrates its superiority in generating razor-shaped fault images over the results produced by other seismic attributes. This attribute, co-rendered with the pseudo relief attribute, clearly demarcates the discontinuous zones of the subsurface with appreciable relief of the prospect (Figure 8.3f). The discontinuous structures within the data volume exhibit large structural variations throughout the data length.

Figure 8.5a shows the example locations of fault-yes and fault-no, which are used for training in all three cases: Cases I (Figure 8.5b), II (Figure 8.5c), and III (Figure 8.5d). We observe that the nRMS errors for Case-I follow the exponential decay for both the train and test data sets (Figure 8.6a). Further, they tend to move parallel and separate out after 10 iterations, resulting in a minimum error of 0.75 and 0.8, and minimum misclassification % of 14.11 and 17.86 % for the train and test data sets respectively. For Case-II, we observe that the nRMS errors for both the train and test data follow exponential decay, and tend to move parallel and probably are unable to separate out after 13 iterations, and the training is stopped (Figure 8.6b). This results in a minimum error of 0.6 and 0.65, and a minimum misclassification % of 11.08 and 11.10 % for the train and test data sets, respectively. However, the nRMS errors for both the train and test data in Case-III (Figure 8.6c) follow exponential decay and tend to move parallel and begin to separate out after 25 iterations. Further, the training is stopped, resulting in a minimum error of 0.35 and 0.40, and minimum misclassification % of 8.02 and 9.67 % for the train and test data sets, respectively.

Moreover, it is observed that the training in Case-III demonstrates a steady improvement (both in minimizing nRMS error and misclassification %) compared to either Case-I or Case-II. This suggests that selection of proper attributes plays a crucial role in designing the meta-attribute, aimed at delimiting the subsurface features, e.g., presence of faults and non-faults or any other structural feature. The training is stopped after certain iterations (10 iterations for Case-I; 13 iterations for Case-II; 25 iterations for Case-III) to avoid overtraining that may lead to erroneous results.

Though attributes in Case-I highlight the geological discontinuities, these are not aligned, e.g., the curvature anomaly is observed on the side of faults, whereas coherence and semblance anomalies are observed along the fault plane (Figures 8.3). Such a combination probably results in a noisy and blur FC meta-attribute (Figure 8.7a). Though the FC meta-attribute for Case-II shows

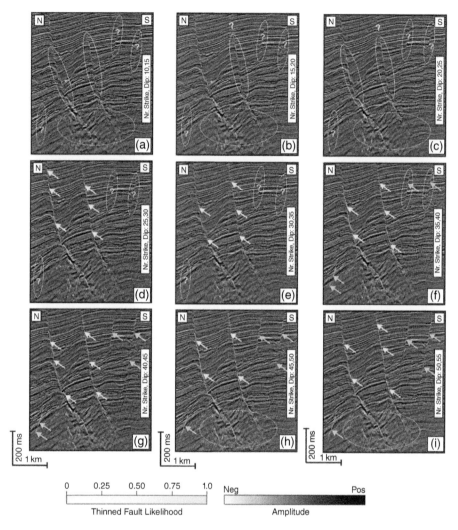

Figure 8.4 (a–i) TFL attributes computed for different scans along the strike and dip orientations (shown near the right boundaries); Scans (a–c) fails to highlight discontinuities at different places. Discontinuities towards southern and deeper parts are poorly imaged (cream dotted oval with question marks). Observations still continue in (d–e). A steady improvement is observed for scans in (f) to (g). Faults in the northern, central and southern part are clearly visible (cream arrows). Minute discontinuities are observed in the deeper part (cream arrows). Scans in (h–i) distinctly bring out structural discontinuities (yellow and cream arrows and cream dotted ovals). Scanning for a wider range helps in capturing maximum structural details. Source: Kumar, P. C., & Sain, K. (2018). Attribute amalgamation-aiding interpretation of faults from seismic data: An example from Waitara 3D prospect in Taranaki basin off New Zealand. *Journal of Applied Geophysics, 159*, 52–68.

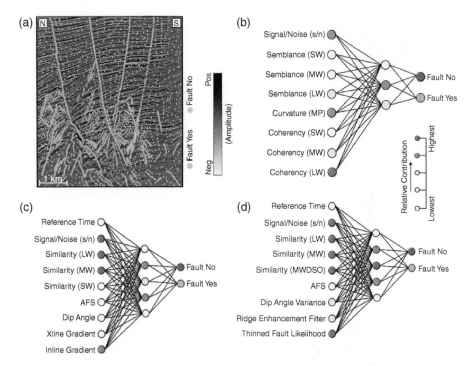

Figure 8.5   (a) Example locations for inline 1078, which are classified into fault-yes (green dots) and fault-no (pink dots) groups. (b) Case-I, (c) Case-II, and (d) Case-III MLP networks. Each layer is interconnected with each other through different nodes by a fully connected network. Color scale pale yellow to red implies relative contribution of each input node to classify the output in terms of fault-yes (green color) and fault-no (pink color). Red color nodes provide the highest contribution to train the network. Source: Kumar, P. C., & Sain, K. (2018). Attribute amalgamation-aiding interpretation of faults from seismic data: An example from Waitara 3D prospect in Taranaki basin off New Zealand. *Journal of Applied Geophysics, 159*, 52–68.

better geologic structures (Figure 8.7b), noises still persist in and around the fault zones. The faults towards the southern part lack in continuity and become smearing. The structural details are also not prominent in the deeper part of the section. However, the FC meta-attribute for Case-III provides thinned, sharp, and razor-shaped images of geologic structures (Figure 8.7c) without any smearing and noise. The fault planes are continuous and their subsurface dispositions are prominently observed.

The structural details in the deeper part for Case-III are also sharper and discrete in FC attribute (Figure 8.7c) as compared to the FC attributes, observed for Case-I (Figure 8.7a) and Case-II (Figure 8.7b), respectively. The time slice at t = 0.88 s, shown over a representative section (Figure 8.8a) for Case-III, further adds value to these observations and interpretation. Though geologic structures

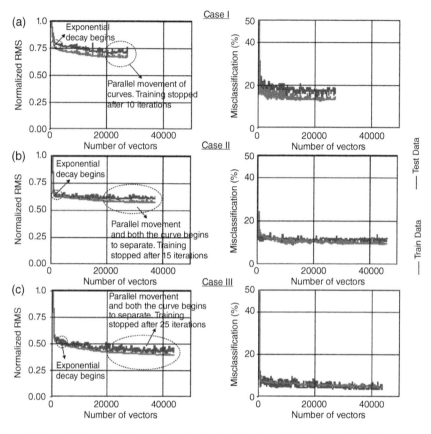

Figure 8.6   nRMS error (left panel) and misclassification % (right panel) for the train (red) and test (blue) data sets, respectively, for (a) Case-I, (b) Case-II, and (c) Case-III, respectively.

are captured by the FC meta-attribute for Case-I (Figure 8.8b), the features are devoid of prominent illumination. Features in the eastern part and faults in the western part lack in continuity. The geologic structures in the south western part are prone to noise and the structural details in the southern and north eastern parts are poorly mapped (Figure 8.8b). The FC meta-attribute for Case-II (Figure 8.8c) demonstrates an improvement in structural features. Nevertheless, the structural details in the southern and north eastern parts are captured by this FC meta-attribute; the structural features are, however, masked by noisy events and many overlapped features. The FC meta-attribute for Case-III (Figure 8.8d) has clearly brought out the best possible images, with continuity and connectivity of geologic structures both in the eastern and western part. Again, the result of FC meta-attribute is free from noises and overlapped features in the north eastern, southern, and south western parts of the study area.

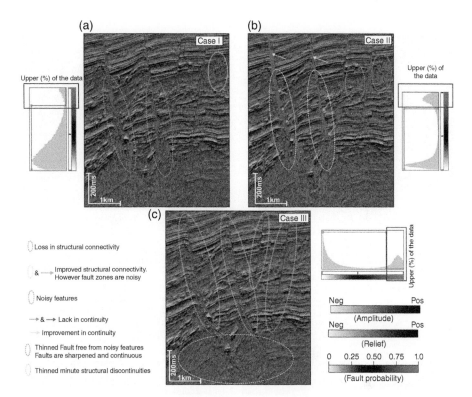

Figure 8.7   FC attribute, co-rendered with amplitude along the seismic inline 1078 for (a) Case-I though showing geological discontinuities but lacks in structural connectivity (orange and pink arrows and orange dotted oval); (b) Case-II showing geological discontinuities with a steady improvement in structural connectivity and continuity (yellow arrows and yellow dotted oval) with some blur in the deeper parts; (c) Case-III exposing images of thin geological discontinuities (blue dotted oval) with distinct structural features and minute structural discontinuities (cream dotted oval). Source: Kumar, P. C., & Sain, K. (2018). Attribute amalgamation-aiding interpretation of faults from seismic data: An example from Waitara 3D prospect in Taranaki basin off New Zealand. *Journal of Applied Geophysics, 159*, 52–68.

Moreover, the FC meta-attribute shows the thinned and sharpened images of the geologic features. The attribute amalgamation in Case-III is thus best suited for the FC meta-attribute generation compared to the FC meta-attributes for either Case-I or Case-II. This demonstrates that a suitable combination of seismic attributes may lead to generation of a meta-attribute that can be used for efficient interpretation of subsurface geologic features from surface data.

To appreciate the geologic trend and orientation of the subsurface faults, the discontinuities (~150 samples) are traced from seismic volume for

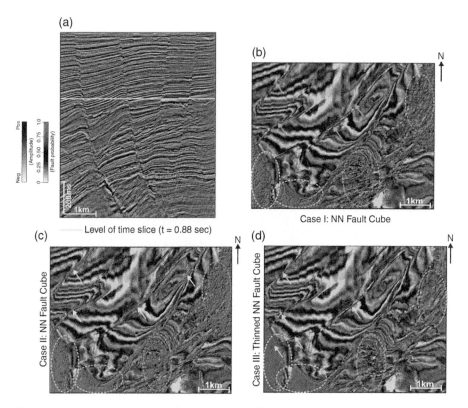

Figure 8.8　(a) FEF time-migrated seismic section for inline 1078 with an yellow line at t = 0.88 s. Time slices at 0.88 s for (b) Case-I, (c) Case-II, and (d) Case-III of FC attribute, co-rendered with the amplitude data. Case-III reveals enhanced structural features compared to those in Case-I and Case-II. The south western areas are devoid of noisy features that are present both in Case-I and Case-II (cream dotted oval). The Case-III meta-attribute thus captures the subsurface structural details more efficiently. Source: Kumar, P. C., & Sain, K. (2018). Attribute amalgamation-aiding interpretation of faults from seismic data: An example from Waitara 3D prospect in Taranaki basin off New Zealand. *Journal of Applied Geophysics, 159*, 52–68.

Case-III at t = 1.2s. Most of the features associated with the faults and discontinuities run parallel and depict NE–SW trends within the prospect (Figure 8.9a). It is observed that geologic discontinuities exhibit two major trends, e.g., N45°E and N60°E (Figure 8.9b). The frequency histograms (Figure 8.9c) reveal that ~80 geologic faults are mostly oriented in between N45°–60°E followed by ~31 and ~41 geologic faults oriented in between N30°–45°E and N60°–75°E, respectively.

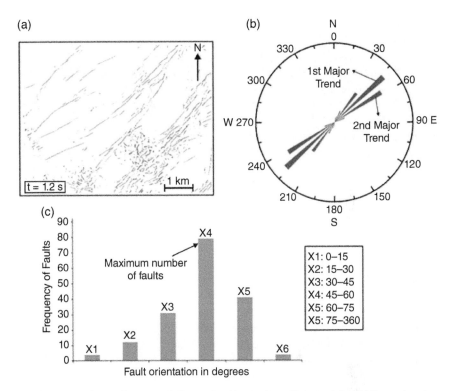

Figure 8.9   (a) Thinned structural discontinuities at time slice, t = 1.2s. (b) Rose plot shows the trend of geologic discontinuities, as observed in (a). Faults and associated discontinuities exhibit two major structural trends, e.g., N45°E and N60°E. (c) Histogram of frequency distribution of fault orientations reveal that most of the faults are oriented between N45°–60°E. (For computing fault orientations, 150 samples have been traced from the thinned display of geologic structures in (a).)

8.4. Efficiency of the Optimized TFC

Though the computation of multi-attributes has been demonstrated by several authors (de Rooij & Tingdahl, 2002; Huang et al., 2017; Kluesner & Brothers, 2016; Kumar & Mandal, 2017; Tingdahl & de Rooij, 2005; Zheng et al., 2014) in preparing the FC meta-attribute, the present discussion is mainly focused on generating a Thinned Fault Cube or TFC meta-attribute using the optimized TFL attribute (Hale, 2013; Wu & Hale, 2016) by scanning the data through several steps within a range of strike 0°–360°) and dip (35°–85°) Directions. The seismic attributes grouped within Cases I, II, and III are extracted from the FEF seismic volume, and the Case-III defines an optimized TFL.

Kumar and Sain (2018) defined this attribute as the Thinned Fault Cube or TFC meta-attribute.

Hale (2013) and Wu and Hale (2016) demonstrated the approach of thinning fault planes that ultimately results in finer structural details. The optimized TFL single attribute illustrates thinned fault images from seismic volume (Figure 8.10a). The geologic faults in the southern part are poorly captured, and the image lacks in structural continuity and connectivity, whereas the TFC meta-attribute in Case-III demonstrates an improved image of geologic features (Figure 8.10b), in which the faults in the southern part are continuous and prominent. Moreover, the TFC meta-attribute has brought out the minute structural details in deeper part of the seismic section. Hence, the fusion of seismic attributes and computing TFC meta-attribute definitely adds value over single attribute (TFL) computation with regard to generating the maximum likelihood of discontinuous structures and faults from seismic volume.

The TFC is very capable compared to a single attribute, e.g., coherency (Bahorich & Farmer, 1995; Marfurt et al., 1998) or similarity (Tingdahl, 2003; Tingdahl & de Groot, 2003), as is evident from Figure 8.11. It can be seen that the coherency attribute delivers smeared images of geologic faults (Figure 8.11a). The structural details are poorly imaged and geologic structures are associated with several noisy and overlapped features that make blurred images of faults. The similarity attribute (Figure 8.11b) improves the fault images compared to the coherency attribute. However, the TFC meta-attribute generates a smooth, sharp, and thinned image of structural discontinuities (Figure 8.11c). The signals along the faulted structures are enhanced, noisy events are suppressed, and structural details are prominently imaged.

Three cases, discussed in the previous paragraphs, highlight the importance of attribute selection and their proper computation in obtaining improved images of geologic structures. Either Case-I, Case-II, or Case-III may be accepted. There may be more case(s) that may be evolved by several other advanced attributes amalgamated over the interpreter's perception. However, the primary concern of an interpreter should be to consistently improve images of geological discontinuities such that several hidden targets are illuminated, noisy or overlapped features are suppressed, and a robust interpretation is achieved. This happens when structurally conditioned seismic data is available for extracting suitable seismic attributes, as has been demonstrated by the three Cases (Figures 8.7 and 8.8), where the approach intends to improve the interpretation of geological discontinuities by making a suitable combination of seismic attributes. When an efficient combination of attributes, which acts as a cornerstone, is used for training over the interpreter's acquaintances, a robust meta-attribute can be generated to aid improved interpretation of seismic data. Moreover, it is obvious that algorithmic differences in the development of attributes

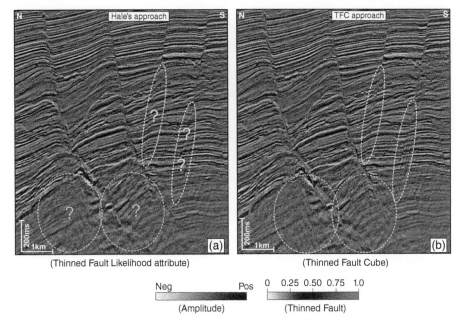

Figure 8.10   (a) Thinned Fault Likelihood (TFL) attribute (single attribute), obtained from Hale's approach and co-rendered with amplitude data for seismic inline 1078, showing thinned fault images and minute structural discontinuities. However, the attribute fails to capture the structural continuity of faults in the southern part (yellow dotted oval with question mark). (b) Thinned Fault Cube (TFC) meta-attribute has captured entire discontinuities, geologic features and faults, and exhibits the efficacy of attribute amalgamation over a single attribute interpretation. Source: Kumar, P. C., & Sain, K. (2018). Attribute amalgamation-aiding interpretation of faults from seismic data: An example from Waitara 3D prospect in Taranaki basin off New Zealand. *Journal of Applied Geophysics, 159*, 52–68.

not only help in improving the images of discontinuity, but also lead to flawless interpretation of seismic data.

8.5. Summary

Major conclusions are drawn from the discussion:
- Though an attribute individually has an ability to capture seismic response of a geologic target (such as a fault), it cannot differentiate similar responses from other geologic target(s).
- The combination of suitable attributes not only reduces the uncertainty while interpreting the subsurface target or feature, but also improves the subsurface image.

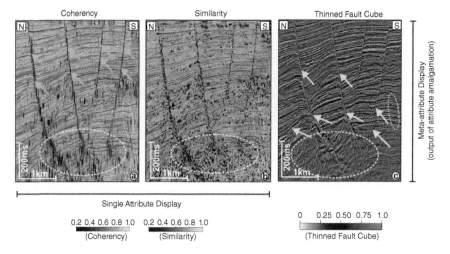

Figure 8.11 (a) Coherency attribute, co-rendered with amplitude data for inline 1077, showing geological discontinuities (green arrows). (b) Similarity attribute shows an improvement of discontinuities (orange arrows). Zones surrounding the faults are still noisy; deeper part is poorly imaged. (c) The TFC meta-attribute shows better images of geologic structures. Faults (yellow arrows) and surrounding areas are noise-free and distinct. Discontinuous structure (orange dotted oval) in the southern part is clearly visible. Structures (yellow dotted oval) in the deeper part are sharpened and finer structural details are prominently captured. Source: Kumar, P. C., & Sain, K. (2018). Attribute amalgamation-aiding interpretation of faults from seismic data: An example from Waitara 3D prospect in Taranaki basin off New Zealand. *Journal of Applied Geophysics, 159,* 52–68.

- A robust interpretation of subsurface geologic features like faults is possible through computation of a meta-attribute by combining multiple other attributes.
- 3D seismic data from Waitara prospect in the Taranaki Basin off New Zealand, which has undergone severe tectonic activities leading to the development of numerous geologic features such as faults, have been used.
- Amalgamation of a set of seismic attributes has led to an optimized TFC meta-attribute that has efficiently captured the subsurface faults, fractures, and associated discontinuities, and illuminated their structural architecture from seismic volume.
- The subsurface features (faults and associated discontinuities) are oriented mostly in N45°E and N60°E.

References

Al-Dossary, S., & Marfurt, K. J. (2006). 3-D volumetric multi-spectral estimates of reflector curvature and rotation. *Geophysics*, *71*, 41–51. https://doi.org/10.1190/1.2242449

Atakulreka, A., & Sutivong, D. (2007). *Avoiding local minima in feedforward neural networks by simultaneous learning.* Paper presented at 20th Australian Joint Conference on Artificial Intelligence. https://doi.org/10.1007/978-3-540-76928-6_12

Bahorich, M., & Farmer, S. (1995). 3-D seismic discontinuity for faults and stratigraphic features: The coherence cube. *Leading Edge*, *14*, 1053–1058. https://doi.org/10.1190/1.1437077

Barnes, A. E. (2003). Shaded relief seismic attribute. *Geophysics*, *68*, 1281–1285. https://doi.org/10.1190/1.1817131

Brouwer, F., & Huck, A. (2011). *An integrated workflow to optimize discontinuity attributes for the imaging of faults*, in 2011 Proceedings: Attributes: New Views on Seismic Imaging—Their Use in Exploration and Production, GCSSEPM, 31st Annual Conference, 496–533.

Chopra, S., & Marfurt, K. J. (2007). *Seismic attributes for prospect identification and reservoir characterization.* SEG, Tulsa. https://doi.org/10.1190/1.9781560801900

Dalley, R. M. (2008). Value of visual attributes: revisiting dip and azimuth displays for 3D seismic interpretation. *First Break*, *26*, 87–91. https://doi.org/10.3997/1365-2397.26.1118.27951

de Groot, P. F. M. (1995). Seismic reservoir characterisation employing factual and simulating wells [Doctoral thesis, TU Delft]. Delft University Press. http://resolver.tudelft.nl/uuid:a596871e-daf9-4ac6-a519-dec802151162

de Groot, P. F. M. (1999). Seismic reservoir characterisation using artificial neural networks. Paper presented at 19th Mintrop Seminar, 16–18.

de Rooij, M., & Tingdahl, K. (2002). Meta-attributes—the key to multivolume, multiattribute interpretation. *Leading Edge*, *21*, 1050–1053. https://doi.org/10.1190/1.1518445

Di, H., & Gao, D. (2016). Volumetric extraction of most positive/negative curvature and flexure attributes for improved fracture characterization from 3D seismic data. *Geophysical Prospecting*, *64*, 1454–1468. https://doi.org/10.1111/1365-2478.12350

Gersztenkorn, A., & Marfurt, K. J. (1999). Eigen structure-based coherence computations as an aid to 3-D structural and stratigraphic mapping. *Geophysics*, *64*, 1468–1479. https://doi.org/10.1190/1.1444651

Hale, D. (2013). Methods to compute fault images, extract fault surfaces and estimate fault throws from 3D seismic images. *Geophysics*, *78*(2), O33–O43. https://doi.org/10.1190/geo2012-0331.1

Huang, L., Dong, X., & Clee, T. E. (2017). A scalable deep learning platform for identifying geologic features from seismic attributes. *Leading Edge*, *36*(3), 249–256. https://doi.org/10.1190/tle36030249.1

Kluesner, J. W., & Brothers, D. S. (2016). Seismic attribute detection of faults and fluid pathways within an active strike-slip shear zone: New insights from high-resolution 3D P-Cable™ seismic data along the Hosgri Fault, offshore California. *Interpretation*, *4*(1), SB131–SB148. https://doi.org/10.1190/INT-2015-0143.1

Kumar, P. C., & Mandal, A. (2017). Enhancement of fault interpretation using multi-attribute analysis and artificial neural network (ANN) approach: A case study from

Taranaki Basin, New Zealand. *Exploration Geophysics*, *49*(3), 409–424. https://doi.org/10.1071/EG16072

Kumar, P. C., & Sain, K. (2018). Attribute amalgamation-aiding interpretation of faults from seismic data: An example from Waitara 3D prospect in Taranaki basin off New Zealand. *Journal of Applied Geophysics*, *159*, 52–68. https://doi.org/10.1016/j.jappgeo.2018.07.023

Luo, Y., Marhoon, M., Al Dossary, S., & Alfaraj, M. (2002). Edge-preserving smoothing and applications. *Leading Edge*, *21*, 136–158. https://doi.org/10.1190/1.1452603

Marfurt, K. J., Kirlin, R. L., Farmer, S. L., & Bahorich, M. S. (1998). 3D seismic attributes using a semblance-based coherency algorithm. *Geophysics*, *64*, 104–111. https://doi.org/10.1190/1.1444415

Roberts, A. (2001). Curvature attributes and their application to 3-D interpreted horizons. *First Break*, *19*, 85–100. https://doi.org/10.1046/j.0263-5046.2001.00142.x

Rosenblatt, F., 1962. *Principles of neurodynamics: Perceptrons and the theory of brain mechanism*. Spartan Books, 616 pp. https://doi.org/10.2307/1419730

Rumelhart, D. E., Durbin, R., Golden, R., & Chauvin, Y. (1995). Backpropagation: The basic theory. In *Backpropagation: Theory, architectures and applications*, Y. Chauvin and D.E. Rumelhart (Eds.), Lawrence Erlbaum Associates, Hillsdale, NJ, 1–34. https://doi.org/10.4324/9780203763247

Singh, D., Kumar, P. C., & Sain, K. (2016). Interpretation of gas chimney from seismic data using artificial neural network: A study from Maari 3D prospect in the Taranaki basin, New Zealand. *Journal of Natural Gas Science and Engineering*, *36*, 339–357. https://doi.org/10.1016/j.jngse.2016.10.039

Tingdahl, K. M., & de Groot, P. F. (2003). Post-stack dip and azimuth processing. *Journal of Seismic Exploration*, *12*, 113–126.

Tingdahl, K. M., (2003). Improving seismic chimney detection using directional attributes. In M. Nikravesh, F. Aminzadeh, L.A. Zadeh (Eds.), *Soft computing and intelligent data analysis in oil exploration. Developments in petroleum science* (Vol. *51*, pp. 157–173). Elsevier.

Tingdahl, K. M. & de Rooij, M. (2005). Semi-automatic detection of faults in 3D seismic data. *Geophysical Prospecting*, *53*, 533–542. https://doi.org/10.1111/j.1365-2478.2005.00489.x

Todd Energy Ltd, 2014. PEP 51558 Interpretation of the Waitara 3D Seismic Survey, NZP&M, Ministry of Business, Innovation & Employment (MBIE), New Zealand. Unpublished Petroleum Report PR4845. 1–12.

Wu, X. (2017). Directional structure-tensor-based coherence to detect seismic faults and channels. Geophysics, 82(2), A13-A17.Wu, X., & Zhu, Z. (2017). Methods to enhance seismic faults and construct fault surfaces. *Computers & Geosciences*, *107*, 37–48. https://doi.org/10.1190/geo2015-0380.1

Wu, X., & Hale, D. (2016). 3D seismic image processing for faults. *Geophysics*, *81*(2), IM1–IM11. https://doi.org/10.1190/geo2016-0473.1

Zheng, Z. H., Kavousi, P., & Di, H. B. (2014). Multi-attributes and neural network-based fault detection in 3D seismic interpretation. In *Advanced Materials Research, Trans Tech Publications* (vol. *838*, pp. 1497–1502). https://doi.org/10.4028/www.scientific.net/AMR.838-841.1497

9

FAULT AND FLUID MIGRATION INTERPRETATION

This chapter shows how two meta-attributes can help to understand the leakage of hydrocarbon fluids through a network of hard-linked normal faults. In a previous chapter, we computed the Thinned Fault Cube meta-attribute and here we demonstrate the computation of the Fluid Cube meta-attribute formed from a suitable combination of seismic attributes through an artificial neural network. The interpretation based on dual meta-attributes showcases how these can delimit the fluid flow pathways through the hard-linked faults observed in the Miocene and Paleocene formations in the study area.

9.1. Introduction

This chapter elucidates fluid leakage through discontinuous geological structures. Two attributes, called the Thinned Fault Cube (TFC) and Fluid Cube (FlC) meta-attributes, are designed from a high-resolution 3-D seismic volume in the Parihaka prospect of Taranaki Basin off New Zealand based on the workflow described in Chapter 8. The geological formations of interest comprise the Tikorangi Formation of Oligocene and the Manganui and Moki Formations of Miocene ages. The chapter starts with a brief description of data used here and then proceeds to interpretation.

Meta-Attributes and Artificial Networking: A New Tool for Seismic Interpretation,
Special Publications 76, First Edition. Kalachand Sain and Priyadarshi Chinmoy Kumar.
© 2022 American Geophysical Union. Published 2022 by John Wiley & Sons, Inc.
DOI: 10.1002/9781119481874.ch09

9.2. Geophysical Data

The study uses 3-D post-stack time-migrated seismic data consisting of 1,132 inlines (Line no. 1,665 to 2,797) and 2,904 xlines (Line no. 2,835 to 5,739) acquired over the Parihaka prospect of the Taranaki Basin (Figure 9.1). The seismic data was acquired in 2005 by B/V Veritas Viking II for Pogo New Zealand covering an area of 1,028.55 km^2. Additional acquisition parameters include a bin spacing of 25.0 m × 12.5m (inl/xl), a 4 ms sampling interval, a 60-fold coverage, and a recording length of 6.0 s. The acquired data have been processed following conventional workflows that include trace edits, geometry updates, signal deconvolution, anomalous amplitude attenuation (AAA), multiple elimination, velocity analysis, dip move-out corrections, migration preconditioning, anisotropic Kirchhoff time migration, and common midpoint (CMP) data stacking. A three-pass velocity analysis, accompanied by a dense velocity analysis (DVA), was applied to the data. The aim of the robust processing sequence was to image deep geological structures (particularly Cretaceous strata), improve fault resolution, and reduce multiple and fault shadow effects (Western Geco, 2012). The seismic volume is displayed following the Society of Exploration Geophysicist's (SEG) American polarity, i.e. an increase in acoustic impedance is shown in black and related to a peak (or positive amplitude), while decreases in acoustic impedance are shown as red reflections and correlated with the troughs, or negative amplitudes. Additional geophysical data in the study area are obtained from exploration wells Arawa-1, Kanuka-1, Okoki-1, Taimana-1, and Witiora-1. The information from Arawa-1 well, which was drilled up to a total depth (TD) of 3,055 m to the top of the Moki Formation, has been utilized (ARCO Petroleum, 1992).

9.3. Results and Interpretation

9.3.1. Thinned Fault Cube (TFC) and Fluid Cube (FIC)

Time-migrated 3-D seismic data show that geological structures are associated with noises and distorted reflections, leading to poor subsurface images (Figure 9.2a, b). Conditioning of such data using the DSMF smoothes the reflectors by reducing the noises, and enhances the continuity of seismic events (Figure 9.2b). Furthermore, application of FEF to the data has sharpened the structures and refined the signal near the fault zones. Fault edges are distinctly and clearly observed (Figure 9.2c).

Seismic attributes are computed from the conditioned data (Figures 9.3 and 9.4), which are complex in nature. The Thinned Fault Likelihood (TFL)

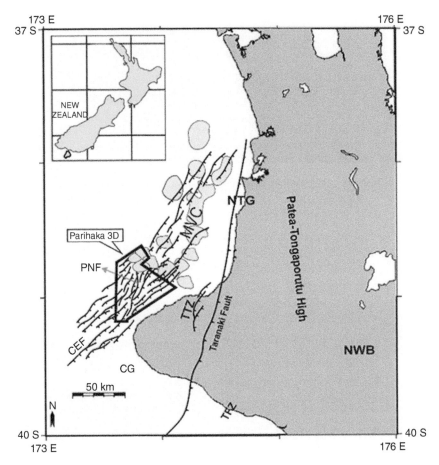

Figure 9.1 Location of the Parihaka prospect in offshore northern Taranaki Basin, New Zealand. The prospect is intersected by extensional faults trending NE-SW. To the NE the prospect is surrounded by buried intrusive bodies forming the Mohakatino Volcanic Centre (MVC). PNF: Parihaka Normal Fault; NTG: Northern Taranaki Graben; CG: Central Graben; TTZ: Tarata Thrust Zone; NWB: Northern Wanganui Basin; CEF: Cape Egmont Fault; TFZ: Taranaki Fault Zone.

attribute (Figure 9.3a) improves the subsurface features when scanned for maximum likelihood, with several steps in the strike and dip directions (Figure 9.3b–f). This has been well demonstrated in Chapter 8. Every minute structural details are captured by increasing the scan from 11 to 51 in the strike and from 14 to 60 in the dip directions (Figures 9.3b–f). Structural features in the deeper

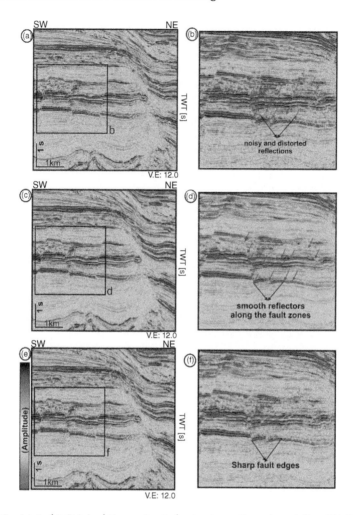

Figure 9.2 (a) & (b) Original time-migrated seismic section along inline (IL) 2231 from Parihaka prospect; fault zones are mixed up with noises and distortions (zoomed view at right panel). (c) & (d) DSMF conditioned section for the same inline with improved fault zones, free from distortions. (e) & (f) FEF applied section exhibiting sharper fault edges and enhanced fault zones. Right panel is the zoomed part of boxes, shown to the left panel.

parts of faults are efficiently apprehended by the TFL attribute (Figure 9.3f). These are ascertained for an optimized case for the TFL attribute (with the number of strike and dip scans: 51, 60) that arrests the maximum number of structural discontinuities from the seismic volume.

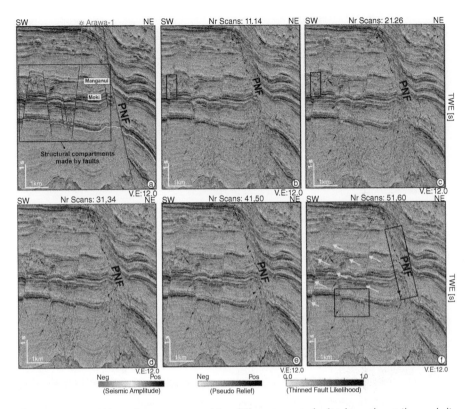

Figure 9.3 (a) TFL attribute computed for different scans (b–f) along the strike and dip orientations; Scans (b–c) fail to highlight discontinuities that are poorly imaged (black rectangle and green arrows) towards south; discontinuities are improved in deeper strata for scans (d–e); faults in the northern, central and southern part are clearly visible (blue arrows). Further scanning (f) brings out structural discontinuities (yellow and orange arrows and black rectangle and square) and maximum structural detail. Parihaka Normal Fault (PNF) is marked on the seismic section.

Attributes such as the similarity and ridge enhancement filter (REF) can improve the discontinuous reflection character of the fault zones (Figures 9.4a, b). The TFL attribute shows linear faults from the data volume (Figure 9.4c). Further, attribute analysis optimally adds value by bringing out fluid migration pathways, as most of these pathways correspond to breached relay ramps. The similarity attribute illuminates contrasting reflection events within these pathways and distinguishes them from surrounding strata (Figure 9.4d). The energy attribute suggests that these zones are associated with low-amplitude and low-energy events (Figure 9.4e). The higher component of frequency is washed out within the fluid migration pathways, as shown by the frequency attribute (Figure 9.4f).

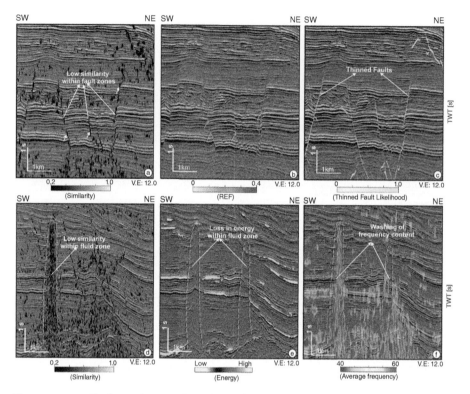

Figure 9.4 (a) Similarity (low similar values), (b) REF (high ridges), and (c) TFL (maximum likelihood of structural disturbance) attributes respond to structural signatures of geologic discontinuities; (d) similarity, (e) energy, and (f) average frequency attributes capture the dissimilar, chaotic and frequency "wash-out" characters of fluid migration zones.

9.3.2. Neural Design for the TFC and FIC

Neural training for the test and train data sets resulted in small nRMS errors and misclassification % (Figures 9.5a–f). The sigmoid activation function takes the input and squashes the output in terms of 0s and 1s, where 0 refers to "fault-no" and 1 refers to "fault-yes" in the case of TFC meta-attribute computation, whereas 0 refers to "fluid-no" and 1 refers to "fluid-yes" for the case of FIC meta-attribute computation. For TFC meta-attribute, the nRMS error and misclassification % for both the train and test data sets reached minimum values of 0.3 and 0.42, and 7.13 and 9.38%, respectively, after 40 iterations (Figure 9.5c). In the case of the FIC meta-attribute, the nRMS error and misclassification % reached minimum values of 0.3 and 0.45, and 3.08 and 7.55%, respectively, for both the train and test data sets, respectively, after 30 iterations (Figure 9.5f). It is important to attain the minimum error between the system-generated

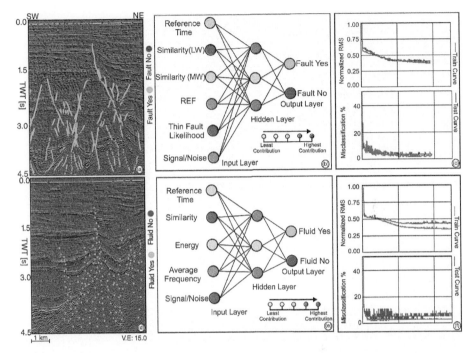

Figure 9.5   (a) Example locations for fault-yes and fault-no zones. (b) Fully connected MLP network used for the computation of TFC meta-attribute. (c) Performance evaluation for constructing the TFC meta-attribute. (d) Example locations for Fluid-Yes and Fluid-No zones. (e) Fully connected MLP network used for the computation of FIC meta-attribute. (f) Performance evaluation for constructing the FIC meta-attribute.

Table 9.1 Relative weight offered by each input node (seismic attribute) for neural training in designing the TFC meta-attribute

Attributes (input nodes)	Weights
TFL	97.2
Similarity LW (Long Window)	93.8
REF	87.2
Signal/Noise	84.5
Similarity MW (Mid Window)	77.3
Reference Time	71.8

response and expected outcome (assigned by interpreter based on properties and characteristics) for both the test and train data sets to ascertain that the network is producing the desired outcome. The relative weights assigned to input seismic attributes are shown in Tables 9.1 and 9.2.

Table 9.2 Relative weight offered by each input node (seismic attribute) for neural training in designing the FIC meta-attribute

Attributes (input nodes)	Weights
Similarity	96.8
Signal/Noise	92.4
Average Frequency	84.8
Energy	74.2
Reference Time	71.8

9.3.3. Interpretation Using TFC and FIC

The TFC meta-attribute brought out thinned continuous faults and structural features of subsurface from the data volume. The Manganui, Moki, and underlying Tikorangi Formations are structurally deformed, and they have been compartmentalized in a series of half-graben structures (Figure 9.6). Faults apprehended by the TFC meta-attribute within the Manganui Formation strike to the NE (Figure 9.7). The master fault i.e., the PNF (Fault F3) presents a sigmoidal geometry and divides the formation into eastern and western structural compartments. The eastern compartment belongs to the hanging-wall domain and the western compartment belongs to the footwall domain (Figure 9.7).

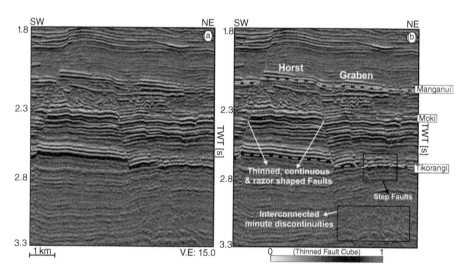

Figure 9.6   (a) Uninterpreted relief attribute, co-rendered with amplitudes along seismic line (IL 2231) exhibiting structural compartments into East and West. (b) Interpreted section of TFC meta-attribute, co-rendered with amplitudes and relief attribute for the same line. The TFC meta-attribute clearly shows thin sharpened faults, which structurally compartmentalize the Miocene (Manganui and Moki) formations and intersect underlying Oligocene (Tikorangi) formation.

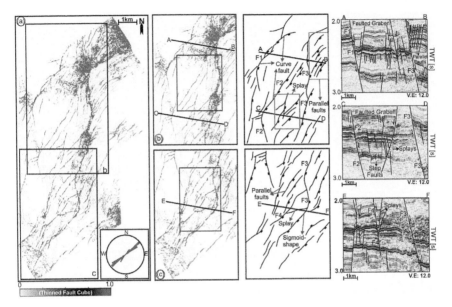

Figure 9.7   (a) TFC meta-attribute displayed at the top of Manganui Formation; faults within the formation demonstrate a NE-SW strike. (b) Faults within the northern to central part show a curved pattern (namely F1 and F2); interpreted faults are shown to immediate right panels; fault splays are observed in the central part of the formation and parallel faults are seen towards the NE part. Fault F3 is segmented and shows a sigmoidal geometry. (c) Faults towards the southern part are Y-shaped (namely, F4); interpreted faults are shown to immediate right. The F4 fault is also associated with several splays. The sigmoid-shape of F3 is more prominent in the southern part. Interpreted faults are shown on seismic sections passing through Lines A-B, C-D, and E-F, as indicated in (b) and (c).

Towards NW of the Manganui Formation, faults exhibit a curved (or semi-circular) geometry (F1 in Figure 9.7b; Table 9.3). These curved faults are further associated with several splays with a horse-tail appearance (Figures 9.7a and 9.8). A similar geometry is also observed in the central part of the Manganui Formation (F2 in Figure 9.7b). The ramps between faults F1-F3 and F2-F3 are hard-linked and breached by several parallel faults and fault splays (Figures 9.7b and 9.8b). Fault F3 is segmented in the northern part, and a sigmoidal shape is more prominent from the center towards the south (Figure 9.7c; Table 9.3). One Y-shaped fault (F4) and associated fault splays are observed to the south (Figures 9.7c, 9.8c, and Table 9.3). These splays breach the ramp formed between F4 and F3 (Figures 9.7c and 9.8c).

The Moki Formation is also deformed by a complex fault system. The faults within the Moki Formation show NNE–SSW strikes (Figure 9.9). Faults in the northern part are curved (F1) and parallel along their strike (Figure 9.9b). In the southern part, a relay ramp between the sigmoidal F3, Y-shaped F4 and

Table 9.3 Identified fault geometry and trends within Miocene strata

Fault ID	Structural architecture	Structural trend	Remarks
F1	Curve shape	N100–120E	Fault contains several splays and breached ramp zones
F2	Curve shape	N110–170E	Fault contains several splays
F3	Sigmoidal/ S-shape	N50–70E	Fault is segmented and separates the whole geological sequence into two major eastern and western structural compartments
F4	Y-shaped	N40–60E	Faults, associated with splays and breached ramps, provide pathways for fluid leakage

curved F2 (Table 9.3) is breached by a series of parallel faults (Figure 9.9c). Discontinuous structures within the underlying Oligocene Tikorangi formation demonstrate two major fault strikes, i.e. NNE–SSW and NE–SW (Figure 9.10). Curved faults (F1 and F2) are prominent and well developed within this latter unit (Figure 9.10a; Table 9.3). In addition, the segmented F3 continues to reveal an s-shaped pattern (Figure 9.10b–c). The ramps between faults F1, F2, and F3 are breached by en echelon faults (step faults) in the northern to central parts. It is observed that the southern part is deformed by several fault splays with Y-shaped geometries (Figure 9.10c; Table 9.3).

Relay ramps within the western structural compartment of the Miocene units are deformed internally, and further breached by faults and minute structural discontinuities (Figures 9.7, 9.8, 9.9, 9.10, and 9.11). Such intense breaching of the relay ramps provides conductive pathways for fluid migrating upwards into the younger strata (Figure 9.11a). The FlC meta-attribute shows these leakage pathways where fluids are observed to migrate through the Miocene succession to younger strata through the hard-linked fault zones (Figure 9.11b–c). Breaching of the relay ramps and fluid leakage through these structures are observed in Figure 9.11. This also exhibits structural features at t = 2.2 s, through which a time slice has been generated (Figure 9.12). It can be observed from Figure 9.11b that the relay ramp between faults F2 and F3 is breached by several parallel faults and smaller structures in the northern part.

To the south, the relay ramp between faults F3 and F4 is breached by several fault splays (Figure 9.12c). These zones are associated with high fluid probability and signify fluid leakage through breached structures (Figures 9.12d–f and 9.13).

The TFL attribute is efficient in generating thinned fault images and its success depends on the number of scans that can bring out subsurface structural details (Hale, 2013). The optimized TFL attribute illustrates thinned fault images from the seismic volume. However, geological structures in the NE part of prospect are poorly captured, and lack in structural continuity and connectivity (Figure 9.3). The TFC meta-attribute improves the interpretation by

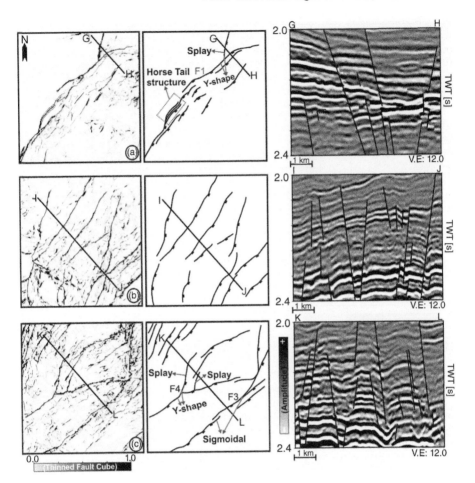

Figure 9.8 TFC meta-attribute displayed for a time slice at t = 2.1 s through the seismic cube at different areas (a, b, c) in the Parihaka prospect; interpreted fault architecture is shown at immediate right panels. (a) Fault F1 is curved in nature and has given rise to several splays arranged in a y-shape pattern. Towards NW, this fault is connected to several other faults arranged in a horse-tail pattern. (b) Central part is deformed with several parallel faults. (c) Fault F4 exhibits Y-shaped structure and has several interconnected splays. Fault F3 is segmented and exhibits a sigmoidal geometry. Interpreted faults are shown on seismic sections at RHS passing through Lines G-H, I-J, and K-L, as shown in (a), (b), and (c), respectively.

capturing the geological features (Figure 9.14). Faults throughout the Parihaka prospect are continuous and prominently observed. Therefore, such a process of amalgamating seismic attributes and generating a hybrid- or meta-attribute is superior in interpreting subsurface feature over a single attribute computation.

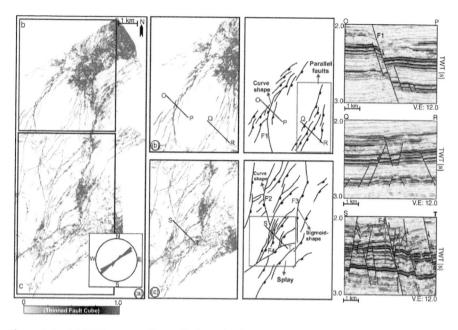

Figure 9.9   (a) TFC meta-attribute displayed at the top of Moki Formation; faults within the formation show a NE-SW strike. (b) Faults within the northern part are curved (namely F1) and parallel (red rectangular box); interpreted faults are shown at immediate right panel. (c) Fault F3 is segmented and exhibits sigmoidal geometry; interpreted faults are shown at the immediate right panel. Faults towards the southern part are Y-shaped (namely, F4). The F4 fault is also associated with several splays (red rectangular box). Interpreted faults are shown on seismic sections at RHS passing through Lines O-P, Q-R, and S-T, as shown in (b) and (c), respectively.

The outcome of TFC meta-attribute is compared with individual attributes, e.g., the coherency attribute of Bahorich & Farmer (1995) and Marfurt et al. (1998) and the similarity attribute of Tingdahl & de Groot (2003) in Figure 9.15. At t = 1.7s, the TFC meta-attribute demonstrates its dominance over the individual coherency and similarity seismic attributes. The coherency and similarity attributes show the geometry of faults but local structural patterns are poorly visible and lack in continuity and connectivity (e.g., the northern zones with yellow square and question marks and the southern zones with red square (Figure 9.15a, b)). The TFC meta-attribute improves the visibility of structures and their continuity (Figure 15c). The advantage of TFC meta-attribute over the individual coherency and similarity attributes is also observed for the time slice at t = 2.0 s (Figure 9.15d, e, f).

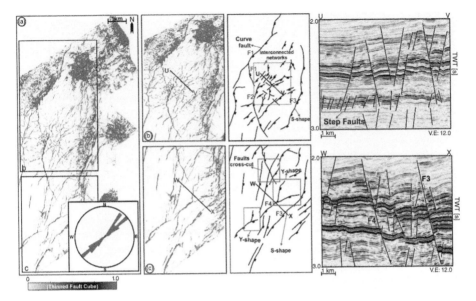

Figure 9.10   (a) TFC meta-attribute displayed at the top of Tikorangi Formation; faults within the formation show a NE-SW strike. (b) Faults within the northern part are curved (F1); interpreted faults are shown at immediate right panel. Central part of the formation is deformed with a set of closely spaced step faults (en echelon faults). (c) Faults towards the southern part are Y-shaped (red rectangles); interpreted faults are shown at immediate right panel. Fault F4 is Y-shaped and associated with several splays. Fault F3 reveals its sigmoidal shape. Interpreted faults are shown on seismic sections at RHS passing through Lines U-V and W-X, as shown in (b) and (c), respectively.

9.4. Summary

The new attributes, called the TFC and FIC meta-attributes, are efficient in capturing faults and fluid migration pathways from 3D seismic reflection data. The new approach has brought out the structural architecture of faults and compartmentalized the Miocene strata of the Parihaka prospect off New Zealand. The main conclusions are:

- A workflow has been designed for the computation of the FIC meta-attribute that has been used in identifying fluid flow pathways through hard-linked faults in the Miocene and Pliocene formations.
- A combination of TFC and FIC meta-attributes is more efficient in delimiting the architecture of hard-linked faults coupled with fluid migration paths to ease the interpretation of subsurface geologic complexities.
- There is a scope for modifying the workflow by incorporating new seismic attributes that may be evolved in future and hence may enhance the interpretation.

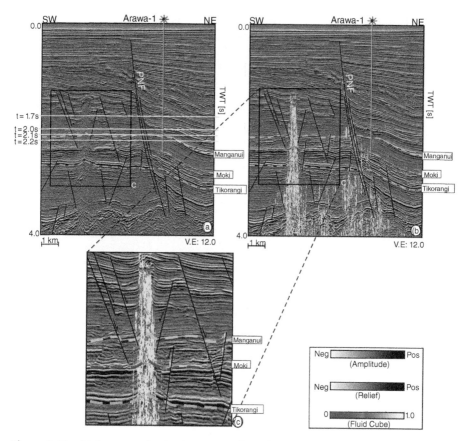

Figure 9.11   (a) Interpreted seismic section along line (IL 1976) showing discontinuous fluid zones within the western compartment of the prospect; the Manganui, Moki, and Tikorangi Formations accommodate these emplaced fluids. (b) The FIC meta-attributes, clipped on seismic section, elucidates vertical migration of fluids through the faulted structures from older (Late Cretaceous-Oligocene) formations to the Miocene intervals, and fluids escape through the younger geological successions. The breached ramps act as conducive pathways for such a leakage. (c) Zoomed view of the fluid zone, as marked by rectangle in (b). PNF: Parihaka Normal Fault, as shown in Figure 9.1. Seismic data, sliced through 2.2 s, as shown in (a), is displayed in Figure 9.12. Scales are shown at the bottom-right corner of Figure. The Arawa-1 well is shown over the seismic section. The horizon or top of Manganui, Moki, and Tikorangi Formations are marked over the seismic section.

- The primary concern of an interpreter should be to consistently improve the image quality such that the responses or seismic attributes from a geologic target are distinct in capturing the target by this semi-automatic approach.

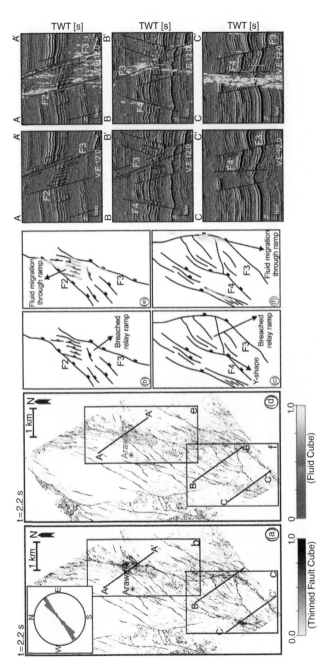

Figure 9.12 (a) Time Slice at t = 2.2 s, displayed through TFC meta-attribute; faults show a NE-SW strike. (b) Ramp between faults F2 and F3 is breached by a chain of parallel faults, thereby deforming the ramp towards NE part of the formation; interpreted faults are shown in the middle panel. (c) The ramp between faults F3 and F4 is breached by parallel faults and splay fault F4, thereby deforming the ramps towards the south; interpreted faults are shown in the middle panel. (d) Time slice at t = 2.2 s, displayed through FIC meta-attribute co-rendered with the TFC meta-attribute. The breached ramps are associated with high FIC probability, suggesting fluid leakage within these hard-linked fault zones; interpreted faults are shown in (e) and (f) in the middle panel. These are also shown on seismic sections at RHS passing through Lines A-A', B-B', and C-C', as shown in (b) and (c) and in (e) and (f), respectively. Scales for TFC and FIC are shown below.

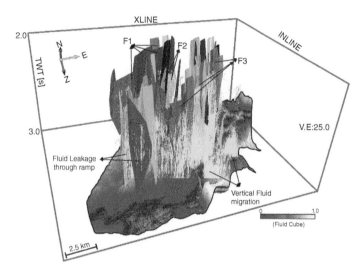

Figure 9.13   3D view of the fluid leakage through the deformed hard-linked zones of the Miocene strata. Fluid leakage is more pronounced in NE-SW trend, where faults are segmented and exhibit complex geometries.

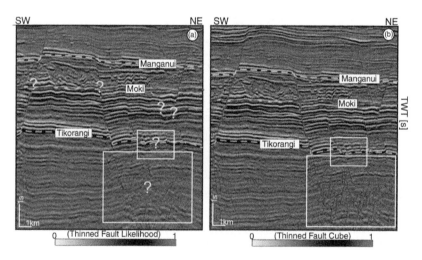

Figure 9.14   (a) TFL attribute, displayed for seismic section along line IL 2231, brings out thinned faults but is poor in capturing structural continuity of faults at some places (indicated by question marks in yellow). Moreover, the TFL attribute fails to bring out minute structural discontinuities in deeper strata. (b) TFC meta-attribute fills these gaps by elucidating minute structural details and faults through different formations (leveled) as well as at the deeper strata, over which the TFC meta-attributes are clipped for display. Scales for TFL and TFC are shown below.

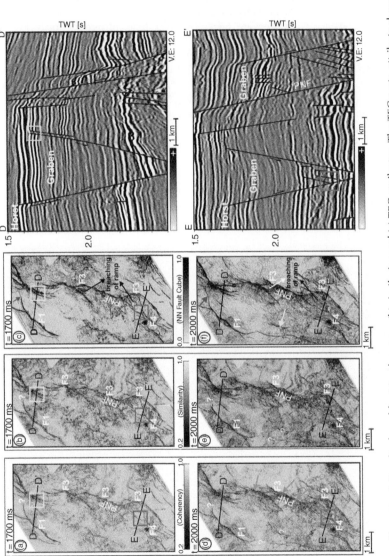

Figure 9.15 Time slice at t = 1.7 s, displayed for (a) coherency, (b) similarity and (c) TFC attributes. The TFC meta-attribute shows enhanced image (c) of faulted structures compared to other images. Time slice at t = 2.0 s, displayed for (d) coherency, (e) similarity, and (f) TFC attributes. Enhancement of subsurface images and minute structural details of faults are again observed for TFC over the individual coherency and similarity attributes. Interpreted faults are shown on seismic sections at RHS passing through lines D-D' and E-E'. The red-colored and yellow-colored rectangle box over lines D-D' and E-E' demonstrates the geometry of fault splays. The comparison between a-b-c and d-e-f highlights the efficacy of meta-attributes over a single attribute interpretation.

References

ARCO Petroleum (1992). Arawa-1 Final well report, *PPL 38447*, Ministry of Economic Development New Zealand: Unpublished Petroleum Report Series *PR 1824*, 1–463.

Bahorich, M., & Farmer, S. (1995). 3-D seismic discontinuity for faults and stratigraphic features: The coherence cube. *Leading Edge, 14*, 1053–1058. https://doi.org/10.1190/1.1437077

Hale, D. (2013). Methods to compute fault images, extract fault surfaces and estimate fault throws from 3D seismic images. *Geophysics, 78*(2), O33–O43. https://doi.org/10.1190/geo2012-0331.1

Marfurt, K. J., Kirlin, R. L., Farmer, S. L., & Bahorich, M. S. (1998). 3D seismic attributes using a semblance-based coherency algorithm. *Geophysics, 64*, 104–111. https://doi.org/10.1190/1.1444415

Tingdahl, K. M., & de Groot, P. F. (2003). Post-stack dip and azimuth processing. *Journal of Seismic Exploration, 12*, 113–126.

Western Geco (2012). Parihaka 3D PSTM Processing report. Ministry of Economic Development New Zealand: Unpublished Petroleum Report Series *PR4582*, 1–135.

10

MAGMATIC SILL INTERPRETATION
(PART 1: TARANAKI BASIN EXAMPLE)

The seismic data have improved our understanding of sill emplacement processes in the subsurface strata. These features are well imaged due to their high-amplitude character, abrupt lateral terminations, and complex geometries within the host rocks. This chapter presents an automated approach to picking up magmatic sills from 3D reflection seismic data with a field example of the offshore Taranaki Basin in the Kora prospect that acts as natural laboratory. It demonstrates how to design the new attribute, called the Sill Cube meta-attribute, from a suitable combination of seismic attributes through an artificial neural network. The study demonstrates the usage of this meta-attribute for delimiting the sill network from 3D seismic data.

10.1. Magmatic Sills: Interpretation Techniques

Seismic reflection methods have been very efficient in studying the sill emplacement processes (Hansen & Cartwright, 2006; Jackson et al., 2013; Magee et al., 2013; Omosanya et al., 2017; Schofield et al., 2012, 2017; Thomson & Hutton, 2004; Thomson & Schofield, 2008). Generally, these features are well-imaged due to their high-amplitude character (which is a function of density and velocity), abrupt lateral terminations, and complex geometries within the host rocks (Alves et al., 2015; Planke et al., 2005; Smallwood & Maresh, 2002). However, 3-D seismic reflection methods suffer from decrease in seismic resolution with depth, overburden effects, and low signal/noise ratio. This

Meta-Attributes and Artificial Networking: A New Tool for Seismic Interpretation,
Special Publications 76, First Edition. Kalachand Sain and Priyadarshi Chinmoy Kumar.
© 2022 American Geophysical Union. Published 2022 by John Wiley & Sons, Inc.
DOI: 10.1002/9781119481874.ch10

creates a problem in distinctively delineating steep dipping and vertical interfaces (Cartwright & Huuse, 2005; Eide et al., 2017; Jackson et al., 2013; Thomson & Schofield, 2008) from the data. Moreover, most of the sills are rarely drilled, which hinders validation.

Several interpretation techniques are available for characterizing sills from seismic data. These include manual interpretation, auto-picking methods, and seismic attribute analysis (e.g., Alves et al., 2015; Cortez & Santos, 2016; Francis, 1982; Hansen & Cartwright, 2006; Jackson et al., 2013; Larsen & Marcussen, 1992; Malthe-Sørensen et al., 2004; Pena et al., 2009; Planke et al., 1999; Smallwood & Maresh, 2002; Thomson & Hutton, 2004; Wood et al., 1988; Zhang et al., 2011). Advanced interpretation based on volume rendering and red-green-blue (RGB) color blending techniques has been a common approach (Alves et al., 2015; Bischoff et al., 2017) in illuminating anomalous features from seismic volumes. The targeted areas are isolated by applying variable opacity values to different amplitudes for extracting relevant geologic features such as the geo-body. The success of these techniques largely depends on the selection of a range of seismic amplitudes that represent the targeted object and an optimum color scale that embodies the ranges such that the surrounding country rock becomes transparent.

A common deficiency for single attribute interpretation is its inability to respond to a specific geologic target within the seismic volume. A possible way to overcome this shortcoming is to compute a meta-attribute from a set of other seismic attributes suited to magmatic sill complexes. This chapter focuses on designing a workflow to compute such a meta-attribute with a view to shedding light on geological implications of a sill network in a complex sedimentary basin of the Kora prospect in the Taranaki Basin off New Zealand (Figure 10.1).

10.2. Research Methods

The workflow adopted to design the seismic Sill Cube (SC) meta-attribute is demonstrated in Figure 10.2. The computation is based on five different steps: (i) structural conditioning of data; (ii) extraction of suitable seismic attributes from data; (iii) selection of example locations; (iv) designing a logical neural network, and (v) validation with other geophysical data and/or well information.

10.2.1. Structural Conditioning

The seismic data volume is first optimally conditioned using a structure-oriented filter (SOF) that utilizes pre-computed dip-azimuth volumes to steer the data in the direction of local dip (Tingdahl, 1999). A dip-steered median filter

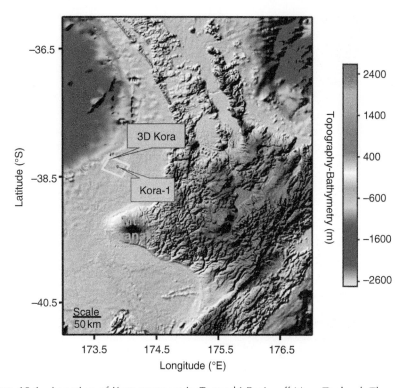

Figure 10.1 Location of Kora prospect in Taranaki Basin off New Zealand. The survey area is demarcated by the yellow rectangle box.

(DSMF) is then applied, which filters the data by applying median statistics over seismic amplitudes following seismic dips. This is performed using a 3×3 median filtering step-out. The conditioned data along with the dip-azimuth steering volume is ready as primary input for the extraction of seismic attributes, followed by designing the ANN architecture and validating the output.

10.2.2. Selection of Attributes

A set of seismic attributes, e.g., amplitude envelope (or reflection strength), texture contrast, entropy, etc., that tend to seize geological responses of sills is then selected. Since several sill complexes are observed at different time levels or two-way time (TWT), an attribute, called "reference time," is also selected

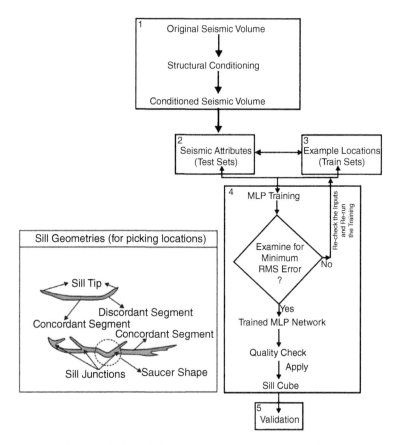

Figure 10.2 Methodological workflow adopted for the present study.

to store the time information from the data. Furthermore, the signal/noise (S/N) ratio attribute is chosen such that the signal level within the sill complexes is maintained.

10.2.3. Example Locations

The example locations are broadly classified into "sill-yes" and "sill-no" classes within the seismic volume. The "sill-yes" class comprises all locations with the presence of sills based on their properties and seismic characteristics, and the "sill-no" class consists of locations without any sill. To have a truly representative set of example locations, an even number of "sill-yes" and "sill-no" locations are chosen, which are assigned 1 and 0 binary codes respectively. Around 3,050 example locations (1,500 sill picks and 1,550 non-sill picks) are selected from the data volume.

10.2.4. Neural Network

A fully connected MLP network is used for designing the meta-attribute. The network consists of three distinct layers: input, hidden, and output layers consisting of 5, 3, and 2 interconnected nodes, respectively. The sigmoid activation function takes input and squashes the output in terms of 0s and 1s, where 0 refers to "sill-no" and 1 refers to "sill-yes." Training of the network begins by randomly splitting a small volume (15-20%) of data into 70% for the train and 30% for the test sets. Iterative training is performed to establish a minimum nRMS error and misclassification % between the observed and predicted outcome. This is also checked by looking into the nRMS and misclassification % for the test data. This ascertains if the network generates the desired outcome. Once this is established, the output is quality checked over a few key seismic lines and the network is run over the entire seismic volume.

10.2.5. Validation

The ability of the designed SC meta-attribute in mapping the sill complexes from the data volume is correlated with the well data and available other information (e.g., Bischoff et al., 2017; Blystad et al., 1995; Omosanya, 2018; Schmiedel et al., 2017; Svensen et al., 2010) for validating the ground truth.

10.3. Results and Interpretation

The 3-D time-migrated seismic volume shows that the sill networks and their overlying folded structures are accompanied by distorted seismic signals and noise (Figure 10.3a). Structural conditioning of the data reduces the noises and enhances the lateral continuity of reflectors that represent the magmatic sill networks and other associated structures (Figure 10.3b).

Most of the sills and associated structures are observed within the Late Cretaceous to Paleocene strata at depths of 3,000–5,000 ms TWT (Figure 10.4a). The emplacement of sills has resulted in the folding of overlying sedimentary layers. The sills exhibit saucer-shaped cross-sectional geometry with their limbs transgressing upwards to the bedding and appear as broken bridge shapes (Figure 10.4b–e). Some sills are also observed as isolated (Figure 10.4c).

We notice that the sills are associated with high reflection strength, high texture contrast, and high entropy as compared to the surrounding strata (Figure 10.5). The sill network shows high reflection strength and entropy both in the western and eastern parts. However, high texture contrast is observed among the sills clustered within the western and central parts of the prospect (Figure 10.5). The neural training is quite satisfactory as evident from the minimum nRMS error and minimum misclassification % between the observed and predicted outcome

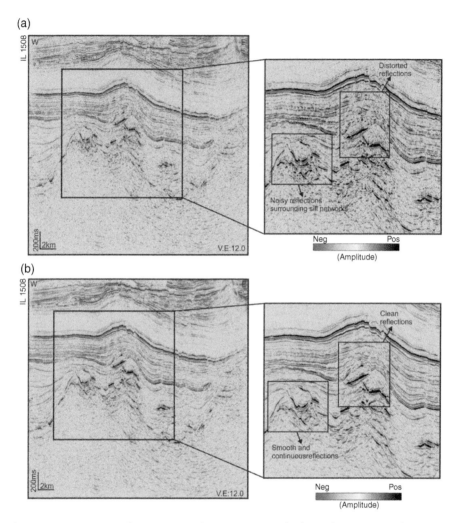

Figure 10.3   (a) Original time-migrated seismic section for line inline (IL) 1508 from Kora prospect. The magmatic sill complexes are mixed up with noises, shown by green arrows at right panel (zoomed view of rectangle in (a)). (b) DSMF conditioned section for the same inline data demonstrating improved signal within the sill complexes, free from distortion, as can be seen at right panel (zoomed view of rectangle in (b)).

for both the train and test data sets (Figure 10.6). The nRMS error of 0.5 and 14.13% and misclassification % of 0.57 and 10.38% are achieved for the train and test data sets, respectively (Figure 10.6a, b, and c) after 30 iterations.

The weights assigned to the inputs seismic attributes are shown in Table 10.1. The extracted output i.e., the SC meta-attribute exhibits interconnected

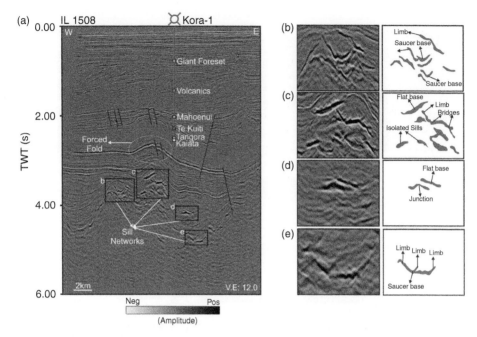

Figure 10.4   (a) Interpreted seismic section along line IL 1508 demonstrating principal formations intersected by the wellbore Kora-1. The subsurface structural elements are faults, folded structures, and sill complexes. The sill networks are individually interpreted to ascertain their geometry and structural architecture. (b) Sills of saucer shape geometry with their limbs concave upwards are interpreted at RHS. (c) & (d) Sills through steps or bridges or junctions, some remain isolated, interpreted at RHS. (e) Sills of saucer geometry with upward rising limbs, interpreted at RHS.

saucer-shaped networks consisting of sills in the deeper part (Figure 10.7a) and enhanced image of sill architecture (Figure 10.7b). This also shows the possible transport pathways of the magmatic fluids into the overlying sedimentary formations.

Forced folds, associated with high SC meta-attributes, are also observed above the sills. This strengthens the fact that these intervals are emplaced by the same magmatic fluids, fed by the underlying sill networks (Figures 10.7b–c). Sills show concave upwards cross-sectional geometry and are connected through steps or bridges (Figures 10.7d–f). The sills are also observed to occur in isolation at t = 4.06 s plan view (Figure 10.8a), and are found interconnected to each other (Figures 10.8b–d). Towards the SW part, isolated sills show saucer-shaped geometry with discordant limbs that are bifurcated, generating sill junctions (Figure 10.8b). On the SE part, the sills are connected through junctions, showing saucer-shaped geometry with rising limbs and fingers through which

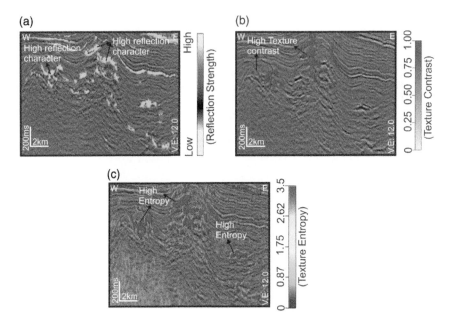

Figure 10.5   (a) Reflection strength attribute showing high reflection character as compared to the surrounding sedimentary units; (b) Texture contrast attribute exhibiting high contrast within the sill networks; (c) Texture entropy attribute displaying the high entropy content within the sill complexes as compared to the surrounding sedimentary environment.

magmatic fluids are emplaced into the surrounding environment (Figure 10.8b). However, the clustered sill networks show broken structures in the NE part (Figure 10.8d).

The isolated sill is estimated to cover an area of ~6.78 sq. km and displays NE–SW trend. On the eastern part, the interconnected sills have NE–SW and NW–SE trends and cover an area of ~12–16 sq. km (Table 10.2). Moreover, the 3-D view of the SC meta-attribute shows the structural architecture of the sill complex that includes limbs, junction, concordant base, fingers, etc. (Figure 10.9). In the deeper part, sills appear to be gently dipping and are interconnected.

10.4. Discussion

10.4.1. Sill Cube: An Efficient Interpretation Tool for Magmatic Sills

Interpretation of magmatic sill complexes becomes tedious when the seismic data is contaminated by coherent and random noises, which lowers the S/N ratio, decreases the accuracy of static and dynamic corrections while processing, and finally degrades the quality of processed sections (Liu et al., 2006; Onajite, 2014).

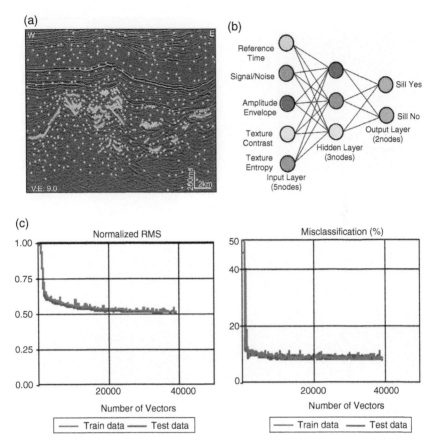

Figure 10.6   (a) Example locations with Sill-yes and Sill-no. (b) Fully connected MLP network. (c) nRMs error and misclassification (%) plots for both the train and test data sets.

Table 10.1 Relative contribution by each input node (seismic attributes) for neural training

Attributes (input nodes)	Weights
Reflection Strength	98.7
Reference Time	79.2
Signal to Noise	86.2
Texture Contrast	78.7
Texture Entropy	85.8

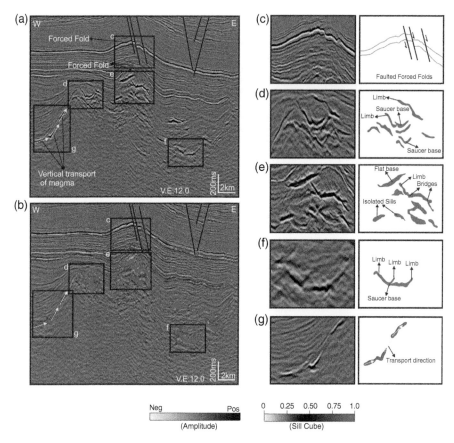

Figure 10.7 (a) Interpreted seismic section showing subsurface architecture of the Kora prospect. (b) Sill Cube (SC) meta-attributes with threshold of 0.75 on seismic amplitudes. (c) Forced folds due to emplacement of sills into the overlying sedimentary strata. (d) Saucer-shaped sill networks depicting concave geometry with their limbs rising upwards. (e) Sills connected through steps or bridges, some are isolated. (f) Sills with saucer geometry and upward rising limbs. (g) Vertical transport of magmatic fluids, indicated by dotted yellow arrows. Interpreted sketches are shown at RHS for the rectangles (c), (d), (e), (f) and (g) as marked in (a) and (b).

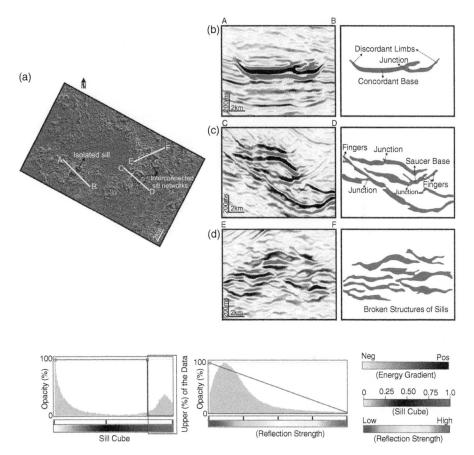

Figure 10.8 (a) Time slice at t = 4.06 s, co-rendered with energy gradient, SC meta-attribute, and reflection strength, show clear picture of magmatic sill complexes. Their presence was further confirmed by displaying seismic sections in (b), (c), and (d) through the lines (AB, CD, and EF) drawn randomly perpendicular to sill trends.

Table 10.2 Area of sills (in km^2) computed using SC meta-attribute in the Kora study area

Sill name	Area of sills (km^2)
Isolated Sill	6.4
Sill SE	15.5
Sill NE	12.0

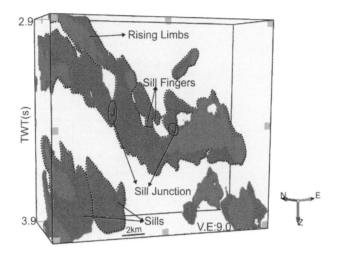

Figure 10.9   3-D view of the sill networks, obtained from the computation of SC meta-attribute. The principal sill with saucer shape geometry and upward rising limb is connected to other sills through junctions and fingers. The limb appears to have step pattern. Several other gently dipping sills are seen below the principal sill.

Thus, it needs to improve seismic responses of magmatic sill complexes from the seismic data. Advanced data conditioning approaches can overcome such difficulties. The original time-migrated 3D seismic volume from the Kora prospect shows distorted seismic signals and noisy reflections (Figure 10.3a). The DSMF has been able to remove the noise bursts that surround the sill networks. The filter effect has been tested through different step-outs (e.g., 3×3; 5×5; 7×7). Over-smoothing of amplitude data is avoided. The application of a mild median filtering step-out (i.e., 3×3) to the seismic volume makes the amplitude content throughout the sill complexes laterally continuous, random noises get suppressed, and S/N ratio is enhanced. This, in turn, increases the visibility of sill networks (associated with high amplitudes) from the surrounding environment in the conditioned data, which does not only preserve the structural architecture but also improves the signal quality (S/N) in and around the sill networks (Figure 10.2b).

Manual interpretation that includes a careful recognition of high-amplitude sill complexes from seismic section (Hansen & Cartwright, 2006; Jackson et al., 2013; Malthe-Sørensen et al., 2004; Thomson & Hutton, 2004) is successful only when the interpreter can effectively distinguish high-amplitude geologic targets from the host-rock strata and geophysical artefacts. Similarly, analysis of seismic attributes depends largely on effective parameter settings such that they can capture the maximum geological responses from the targets (Alves et al., 2015; Barnes, 2016;

Chopra & Marfurt, 2007). For geo-body extraction or opacity rendering techniques (Alves et al., 2015; Bischoff et al., 2017; Schofield et al., 2012; Smallwood & Maresh, 2002; Thomson & Hutton, 2004), the input attribute is processed within a user-defined window to seize a range of amplitude values corresponding to sill complexes and the surrounding strata. These amplitudes are stored in a volume and then interconnected through a voxel connectivity filter to create continuous bodies. To visualize high-amplitude bodies (i.e., sill networks), an appreciable color scale is used, where the host-rock strata are made transparent against the magmatic sills. The downside of using a single attribute in such an approach is that a seismic attribute on its own hardly ever responds to one geological target of interest, and hence brings out a mixed set of geological responses (de Groot et al., 2004; Meldahl, 2002; Tingdahl, 2003). Barnes (2016) suggested that geo-body extraction could be improved by employing multiple attributes or making use of computer-oriented algorithms such that a noise-free image of the targeted body can be made. All these exercises again become very tiresome when dealing with large volumes of 3D seismic data.

A realistic semi-automatic approach is presented here for the interpretation of magmatic sills from 3D seismic reflection data. All possible locations that mark the presence of sill (i.e., the maximum probability) and all possible locations that are devoid of sill (i.e., the least probability) are selected (Figure 10.6a). Furthermore, this judgment is counterchecked with the responses revealed by multiple seismic attributes. Finally, both the judgment and the responses are linked through a fully connected MLP network to minimize the error so that an optimal output is obtained (Figure 10.6a, b, and c). This, in turn, generates a meta-attribute that delimits the magmatic sills from the data volume (Figures 10.7, 10.8, and 10.9).

10.4.2. Limitations of the Sill Cube Automated Approach

The SC meta-attribute is an automated approach that aims to improve the interpretation strategies of sill complexes from 3D seismic data by exploiting the generalization capability of ANN. The downside of this approach relates to the use of noisy or poor quality data as an input for the SC operation. For example, we have not tested the efficacy of this method to mask and remove artefacts such as acquisition footprints. Using such data may result in feeding inaccurate example locations to the network. This ultimately lowers the learning capability of the network leading to generation of artefacts and furthermore impedes steady interpretation. In such instances, the SC may struggle to deliver a successful interpretation of magmatic sill complexes from the data volume.

10.5. Conclusions

The major conclusions are summarized as:

- Amalgamation of a set of suitable seismic attributes and training over interpreter's familiarity with neural algorithms have yielded the SC meta-attribute.
- The SC meta-attribute has successfully captured the sill networks over the structurally deformed sedimentary formations within the Kora prospect in the Taranaki Basin off New Zealand.
- This approach of interpreting magmatic sills from seismic volume is not painstaking, as it incorporates the interpreter's intelligence into a small volume of data to automate the process of interpretation from a large volume of data within a limited time.
- It is noteworthy that several new attributes may evolve, which could be incorporated into the algorithm to improve the computation of the SC meta-attribute and hence the interpretation of magmatic sills from a 3D data volume.

References

Alves, T. M., Omosanya, K. D., & Gowling, P. (2015). Volume rendering of enigmatic high-amplitude anomalies in southeast Brazil: A workflow to distinguish lithologic features from fluid accumulations. *Interpretation 3*: A1–A14.

Barnes, A. E. (2016). Handbook of Post-stack seismic attributes. SEG, Tulsa.

Bischoff, A. P., Andrew, N., & Beggs, M. (2017). Stratigraphy of architectural elements in a buried volcanic system and implications for hydrocarbon exploration. *Interpretation, 5*, SK141–159.

Blystad, P., Brekke, H., Faerseth, R., Larsen, B., Skogseid, J., & Toradbakken, B. (1995). Structural elements of the Norwegian continental shelf, part II, the Norwegian Sea region. Norwegian Petroleum Directorate, Stavanger Norway Bulletin, 6–45.

Cartwright, J., & Huuse, M. (2005). 3D seismic technology: the geological "Hubble." *Basin Research,17*, 1–20.

Chopra, S., & Marfurt, K. J. (2007). Seismic attributes for prospect identification and reservoir characterization. SEG, Tulsa.

Cortez, M. M., & Santos, M. A. C. (2016). Seismic interpretation, attribute analysis, and illumination study for targets below a volcanic-sedimentary succession, Santos Basin, offshore Brazil. *Interpretation 4*: SB37–SB50.

de Groot, P., Ligtenberg, H., Oldenziel, T., Connolly, D., & Meldahl, P. (2004). Examples of multi-attribute, neural network-based seismic object detection. *Geological Society London, 29*, 335–338.

Eide, C. H., Schofield, N., Lecomte, I., Buckley, S. J., & Howell, J. A. (2017). Seismic interpretation of sill complexes in sedimentary basins: implications for the sub-sill imaging problem. *Journal of the Geological Society, 175*, 193–209.

Francis, E. H. (1982). Magma and sediment-I Emplacement mechanism of late Carboniferous tholeiite sills in northern Britain: President's anniversary address 1981. *Journal of Geological Society, 139*, 1–20.

Hansen, D. M., & Cartwright, J. (2006). Saucer-shaped sill with lobate morphology revealed by 3D seismic data: implications for resolving a shallow-level sill emplacement mechanism. *Journal of Geological Society, 163*, 509–523.

Jackson, C. A., Schofield, N., & Golenkov, B. (2013). Geometry and controls on the development of igneous sill–related forced folds: A 2-D seismic reflection case study from offshore southern Australia. *Geological Society of America Bulletin, 125*, 1874–1890.

Larsen, H. C., & Marcussen, C. (1992). Sill-intrusion, flood basalt emplacement and deep crustal structure of the Scores by Sund region, East Greenland. *Geological Society London, Special Publication, 68*, 365–386.

Liu, C., Liu, Y., Yang, B., Wang, D., & Sun, J. (2006). A 2D multistage median filter to reduce random seismic noise. *Geophysics, 71*, V105–V110.

Magee, C., Hunt-Stewart, E., & Jackson, C. A. L. (2013). Volcano growth mechanisms and the role of sub-volcanic intrusions: Insights from 2D seismic reflection data. *Earth and Planetary Science Letters, 373*, 41–53.

Malthe-Sørenssen, A., Planke, S., Svensen, H., Jamtveit, B., Breitkreuz, C., & Petford, N., (2004). Formation of saucer-shaped sills. Physical Geology of High-Level Magmatic Systems. *Geological Society London, Special Publication, 234*, 215–227.

Meldahl, P., Najjar, N., Oldenziel-Dijkstra, T., & Ligtenberg, H. (2002). Semi-Automated Detection of 4D Anomalies. Paper presented at 64th EAGE Conference & Exhibition.

Omosanya, K. O., Johansen, S. E., Eruteya, O. E., & Waldmann, N. (2017). Forced folding and complex overburden deformation associated with magmatic intrusion in the Vøring Basin, offshore Norway. *Tectonophysics, 706*, 14–34.

Omosanya, K. O. (2018). Episodic fluid flow as a trigger for Miocene-Pliocene slope instability on the Utgard High, *Norwegian Sea. Basin Research.* 1–23.

Onajite, E. (2014). Understanding noise in seismic record. In: *Seismic data analysis techniques in hydrocarbon exploration* (pp. 63–68). Elsevier.

Pena, V., Chávez-Pérez, S., Vázquez-García, M., & Marfurt, K. J. (2009). Impact of shallow volcanics on seismic data quality in Chicontepec Basin, Mexico. *Leading Edge. 28*, 674–679.

Planke, S., Alvestad, E., & Eldholm, O. (1999). Seismic characteristics of basaltic extrusive and intrusive rocks. *Leading Edge. 18*, 342–348.

Planke, S., Rasmussen, T., Rey, S. S., & Myklebust, R. (2005). Seismic characteristics and distribution of volcanic intrusions and hydrothermal vent complexes in the Vøring and Møre basins. *Geological Society London, Petroleum Geological Conference Series*, Geol. Conf. Ser. *6*, 833–844.

Schmiedel, T., Kjoberg, S., Planke, S., Magee, C., Galland, O., Schofield, N., Jackson, C. A. L., & Jerram, D. A. (2017). Mechanisms of overburden deformation associated with the emplacement of the Tulipan sill, mid-Norwegian margin. *Interpretation, 5*, SK23–38.

Schofield, N. J., Brown, D. J., Magee, C., & Stevenson, C. T. (2012). Sill morphology and comparison of brittle and non-brittle emplacement mechanisms. *Journal of Geological Society, 169*, 127–141.

Schofield, N., Holford, S., Millett, J., Brown, D., Jolley, D., Passey, S. R., Muirhead, D., Grove, C., Magee, C., Murray, J., & Hole, M. (2017). Regional magma plumbing and emplacement mechanisms of the Faroe-Shetland Sill Complex: implications for magma transport and petroleum systems within sedimentary basins. *Basin Research. 29*, 41–63.

Smallwood, J. R., & Maresh, J. (2002). The properties, morphology and distribution of igneous sills: modelling, borehole data and 3D seismic from the Faroe-Shetland area. *Geological Society London, Special Publication, 197,* 271–306.

Svensen, H., Planke, S., & Corfu, F. (2010). Zircon dating ties NE Atlantic sill emplacement to initial Eocene global warming. *Journal of Geological Society, 167,* 433–436.

Thomson, K., & Hutton, D. (2004). Geometry and growth of sill complexes: insights using 3D seismic from the North Rockall Trough. Bulletin of Volcanology, *66,* 364–375.

Thomson, K., & Schofield, N. (2008). Lithological and structural controls on the emplacement and morphology of sills in sedimentary basins. *Geological Society London Special Publication, 302,* 31–44.

Tingdahl, K. M. (1999). Improving seismic detectability using intrinsic directionality: Technical Report, B194, Earth Sciences Centre, Goteborg University.

Tingdahl, K. M. (2003). Improving seismic chimney detection using directional attributes. In M. Nikravesh, F. Aminzadeh, L.A. Zadeh (Eds.), *Soft computing and intelligent data analysis in oil exploration. Developments in petroleum science* (Vol. *51,* pp. 157–173). Elsevier.

Wood, M. V., Hall, J., & Doody, J. J. (1988). Distribution of early Tertiary lavas in the NE Rockall Trough. *Geological Society London, Special Publication. 39,* 283–292.

Zhang, K., Marfurt, K. J., Wan, Z., & Zhan, S. (2011). Seismic attribute illumination of an igneous reservoir in China. *Leading Edge. 30,* 266–270.

11

MAGMATIC SILL INTERPRETATION
(PART 2: VØRING BASIN EXAMPLE)

This chapter is complementary to the previous chapter, in which we have demonstrated the computation of the sill cube meta-attribute but in the offshore Vøring Basin. The automatic interpretation of magmatic sill networks in the Utgard 3D prospect and its validation with available information demonstrates the application of this new approach to another geo-tectonic world basin.

11.1. Introduction: The Vøring Basin Case

The presence of magmatic sill complexes in Vøring Basin, offshore Norway, is well documented (Blystad et al., 1995; Omosanya, 2018; Schmiedel et al., 2017; Svensen et al., 2010). This chapter demonstrates the efficacy of the SC meta-attribute for the interpretation of magmatic sill complexes in the Utgard prospect within the Vøring Basin. It starts with a brief description of the data followed by expounding interpretation based on the approach, as described in Chapter 10.

11.2. Description of the Data

The data consists of high-resolution, time-migrated, 3-D seismic reflection data (VGUH0201_3D or 3D Utgard) that were acquired over an area of ~1,800 km^2, located close to the Utgard High (Fig 11.1). This 3-D survey is

Meta-Attributes and Artificial Networking: A New Tool for Seismic Interpretation,
Special Publications 76, First Edition. Kalachand Sain and Priyadarshi Chinmoy Kumar.
© 2022 American Geophysical Union. Published 2022 by John Wiley & Sons, Inc.
DOI: 10.1002/9781119481874.ch11

widely known as the Utgard 3-D prospect. The inline (W–E) and xline (N–S) spacing of the survey is 25 m, and the data has a record length of 9,000 ms two-way time (TWT). Furthermore, the data has a sampling rate of 4 ms, yielding a Nyquist frequency of 125 Hz. With a dominant frequency of 40 Hz and velocities of ~4,400 m/s for the magmatic sills, the limit (λ /4) of vertical resolution (Sheriff & Geldart, 1995) is 28 m. Additionally, two exploration wellbores (6607/5-1 and 6607/5-2) were also used for validation and interpretation of magmatic sills. Wellbore 6607/5-2 penetrated into the sill complexes that were interpreted by several researchers (Blystad et al., 1995; Omosanya, 2018; Svensen et al., 2010).

11.3. Interpretation Based on SC Meta-Attribute Computation

Distorted seismic signals and noises were initially observed around several saucer-shaped sills within the Utgard survey area (Figure 11.2). Conditioning of the data through DSMF enhances the lateral continuity of sills within the

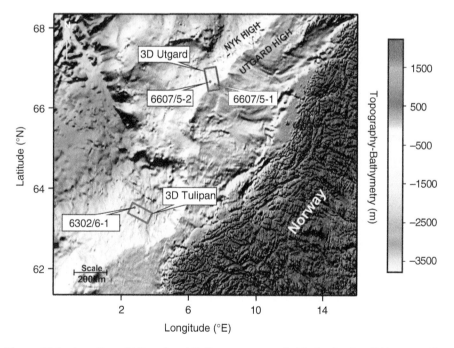

Figure 11.1 Location of Utgard and Tulipan prospects in Vøring basin off Norway. The Utgard prospect (focus of the present study) is drilled by wellbores 6607/5-1 and 6607/5-2. The Tulipan prospect, west of Utgard, is drilled by wellbore 6302/6-1.

Cretaceous to Paleocene strata at depths of 3,000–5,500 ms TWT (Figures 11.2e, 11.2f, and 11.3). The magmatic sills show saucer-shaped geometry with concave upwards morphology (Figure 11.3b) and are interconnected at several junctions (Figure 11.3b–c). The sills in the deeper part exhibit broken bridge structures. Such types of structures occur when overlapping lobes of magma propagate at the same time, generating a bridge between them (Schofield et al., 2012). In vertical sections, these structures are arranged in step patterns, indicating the direction of magma propagation (Figure 11.3c).

The seismic attributes expresses that the sills are associated with high reflection strength, high texture contrast, and high entropy (Figure 11.4). The sills in the northern and southern parts of the survey area have high reflection and high entropy character with saucer shapes (Figure 11.4a, c, e). In the northern part,

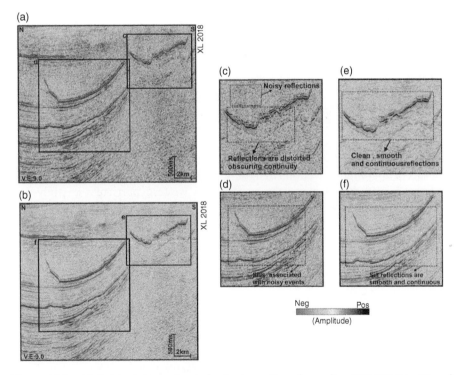

Figure 11.2 (a) Original time-migrated seismic section along xline (XL) 2018 in Utgard area, showing magmatic sill complexes, obscured by noises. (b) DSMF conditioned section along the same line showing improved image within the sill complexes. (c–d) Zoomed view of sill complex, marked by rectangles in (a). (e–f) Zoomed view of sill complexes, marked by rectangles in (b).

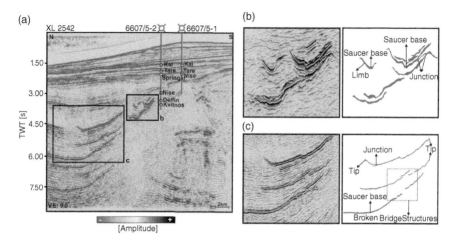

Figure 11.3 (a) Interpreted seismic section along line XL 2542, showing principal formations and structural architecture of magmatic sill complexes intersected by wellbores 6607/5-2 and 6607/5-1. (b) Sill complexes depicting saucer-shaped geometry with their limbs concave upwards and interconnected to each other through junctions, marked by rectangle in (a), immediately right panel shows the interpreted geometry. (c) Sills in the deeper part shows broken bridge structures and step patterns along with saucer geometry, marked by rectangle in (a); panel immediately right shows the interpreted geometry.

high contrast is observed only along the limbs of the sills (Figure 11.4c). However, the sills exhibit high reflection strength, high contrast, and high entropy throughout the saucer shapes along the eastern and western parts of the study area (Figure 11.4b, d, f).

Figure 11.5 shows the example locations for "sill-yes" and "sill-no" and the fully connected MLP network. The sigmoid activation function takes input and squashes the output in terms of 0s and 1s, where 0 refers to "sill-no" and 1 refers to "sill-yes." The neural training achieves a minimum nRMS error of 0.35 and 0.4, and misclassification % of 5.13% and 2.38% after 40 iterations (Figure 11.5c) for both the train and test data sets, respectively. Table 11.1 shows that the amplitude envelope has the highest weight followed by the texture entropy, reference time, S/N ratio, and texture contrast attributes. The SC meta-attribute points out that the sills are interconnected at junctions, saucer-shaped, and have concave upwards geometry with discordant limbs (Figure 11.6). In map view at time slice t = 3.6 s, the sills exhibit saucer-shaped geometry towards the western part, and become clustered with broken or step geometry in the northern part (Figure 11.7). Random cross-sections GH and IJ help in understanding these observations

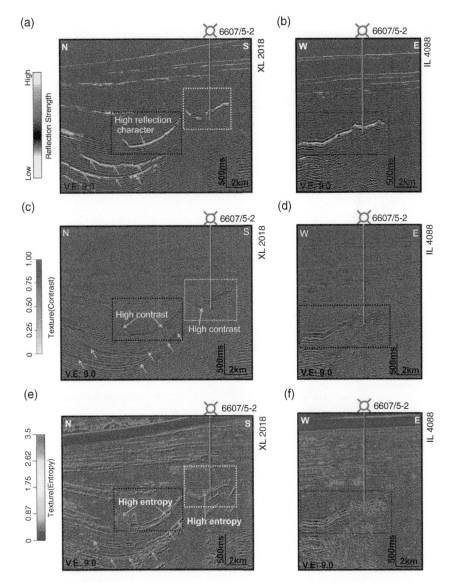

Figure 11.4   Reflection strength attribute along seismic line (a) XL 2018 (N-S profile) and (b) IL 4088 (E-W profile) showing high reflection character (black and green dotted box and orange arrows) as compared to the surroundings sedimentary units. (c–d) Texture attribute for the same lines, displaying high contrast within the sill networks (black and green dotted box and cream arrows). (e–f) Texture entropy attribute for the same lines, showing higher entropy content within the sill complexes compared to the surrounding sedimentary environment (black and green dotted box and blue arrows). Note that wellbore 6607/5-2 has intersected the sill complex.

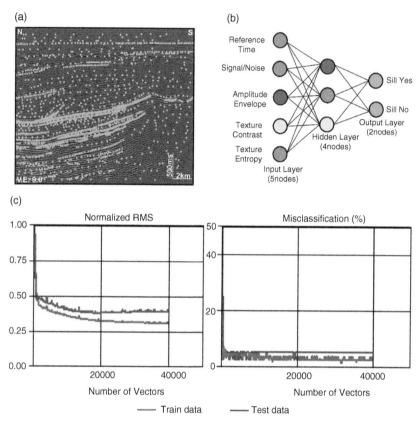

Figure 11.5 (a) Example locations showing Sill-Yes and Sill-No picks for the Utgard study area. (b) Fully connected MLP network used for neural operation. (c) nRMS error and misclassification (%) plots for both the train and test data sets.

Table 11.1 Relative contributions offered by input nodes (seismic attributes) for neural training

Attributes (input nodes)	Weights
Reflection Strength	99.1
Reference Time	84.2
Signal/Noise	86.8
Texture Contrast	62.1
Texture Entropy	75.8

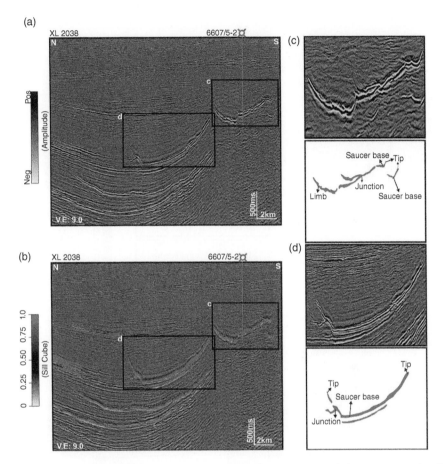

Figure 11.6 (a) Interpreted seismic section along line XL 2038 exhibiting magmatic sill complexes within the Utgard area. The sill network in SE part is intersected by borehole 6607/5-2. (b) SC meta-attribute co-rendered with the interpreted seismic line. (c) Sill network, marked by rectangle in (a–b) shows saucer-shaped structural geometry with limbs rising upwards and interconnected through junctions, interpreted geometry shown at the bottom. (d) Similar structural geometry is also observed for the sill complexes, marked by rectangle in (a–b). Maximum probability of meta-attribute with a threshold of 0.75, indicating sill complex, is clipped over the section.

(Figure 11.7b and c). Overall structural geometry of the sills is displayed in Figure 11.8. Six sills have been interpreted: Sill 1 has a pointed base and Sills 2-6 have concave upwards geometries. Sills 2 and 3 correspond to the "Utgard Upper and Lower Sill" (Svensen et al., 2010). Based on SC meta-attributes, sills in the Utgard area cover an area of 105–1,140 km^2 (Table 11.2).

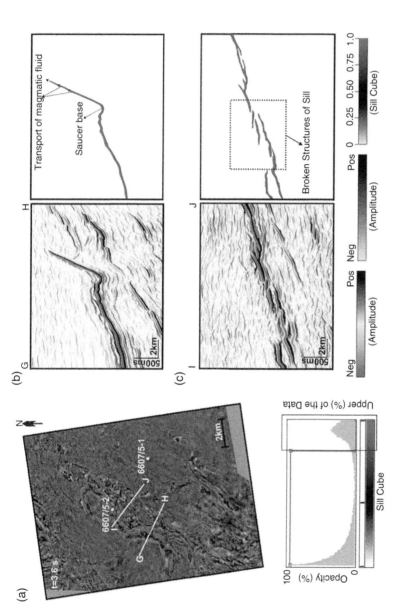

Figure 11.7 (a) Time Slice at t = 3.6 s showing amplitude data, co-rendered with SC meta-attribute. This has brought out clear picture of magmatic sill complexes and their structural geometry. (b) Seismic section along a random line GH as shown in (a) illustrates saucer-shaped geometry of sill with concave upward limb. Magmatic fluids are transported through the limbs (black dotted arrows) and emplaced into the surrounding sedimentary units. (c) Seismic section along IJ another random line as shown in (a) exhibits broken or step geometry of sill network towards northern part of the Utgard area.

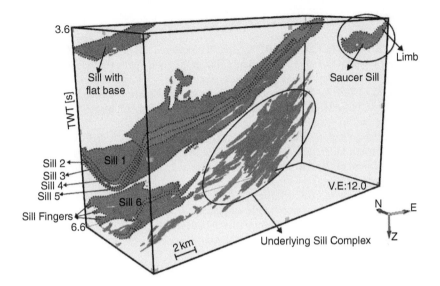

Figure 11.8   3D view of sill networks (in the Utgard area) that has been delineated by SC meta-attribute. Most of the sills (Sill 1-6) depict saucer geometry with concave upwards rising limbs. The underlying sill networks are clustered with broken geometry and run parallel to each other.

Table 11.2   Area of sills (in km^2) computed using SC meta-attribute in Utgard study area

Sill name	Area of sills (km^2)
Sill 1	574.5
Sill 2	1139.8
Sill 3	1030
Sill 4	585
Sill 5	290
Sill 6	105.6

11.4. Summary

The main conclusions are:

- Unification of a set of seismic attributes based on ANN algorithms results in a new SC meta-attribute.
- The SC meta-attribute has successfully delimited the geometry of subsurface sill networks in the Utgard prospect of Vøring offshore basin in Norway.
- This aids in semi-automatic interpretation of magmatic sills from 3D seismic data with a reduced need for intervention by human analysts.

References

Blystad, P., Brekke, H., Faerseth, R., Larsen, B., Skogseid, J., & Toradbakken, B. (1995). Structural elements of the Norwegian continental shelf, part II, the Norwegian Sea region. Norwegian Petroleum Directorate, Stavanger Norway Bulletin, 6–45.

Omosanya, K. O. (2018). Episodic fluid flow as a trigger for Miocene-Pliocene slope instability on the Utgard High, Norwegian Sea. *Basin Research. 1–23*. https://doi.org/10.1111/bre.12288

Schmiedel, T., Kjoberg, S., Planke, S., Magee, C., Galland, O., Schofield, N., Jackson, C. A. L., & Jerram, D. A. (2017). Mechanisms of overburden deformation associated with the emplacement of the Tulipan sill, mid-Norwegian margin. *Interpretation, 5*, SK23–38. https://doi.org/10.1190/INT-2016-0155.1

Schofield, N. J., Brown, D. J., Magee, C., & Stevenson, C. T. (2012). Sill morphology and comparison of brittle and non-brittle emplacement mechanisms. *Journal of Geological Society, 169*, 127–141. https://doi.org/10.1144/0016-76492011-078

Sheriff, R. E., & Geldart, L. P. (1995). *Exploration seismology*. Cambridge, UK: Cambridge University Press.

Svensen, H., Planke, S., & Corfu, F. (2010). Zircon dating ties NE Atlantic sill emplacement to initial Eocene global warming. *Journal of Geological Society, 167*, 433–436. https://doi.org/10.1144/0016-76492009-125

12

MAGMATIC SILL AND FLUID PLUMBING INTERPRETATION (CANTERBURY BASIN EXAMPLE)

This chapter shows the workflow for the computation of the fluid cube meta-attribute along with the sill cube meta-attribute, as demonstrated in the previous two chapters. The study explains the sill intrusions and fluxed-out magmatic fluid pathways into the surrounding sedimentary strata in the Waka prospect within the Canterbury basin, offshore New Zealand, by designing the sill and fluid cube meta-attributes by merging appropriate attributes based on artificial neural networks. It is observed that magma routing has caused phreatic activity leading to circulation of superheated fluids that vertically rise to a height of ~800 m within the encased sedimentary strata.

12.1. Introduction: The Canterbury Basin Case

The offshore Canterbury Basin has undergone complex tectonism and remained prone to different stages of volcanism. This has created a center of interest for the exploration of magmatic sill complexes to understand the emplacement processes (Figure 12.1). The present chapter aims to showcase the efficiency of the SC meta-attribute approach (already described in Chapters 10 and 11) for the interpretation of sill intrusions and fluxed-out magmatic fluid pathways into

Meta-Attributes and Artificial Networking: A New Tool for Seismic Interpretation,
Special Publications 76, First Edition. Kalachand Sain and Priyadarshi Chinmoy Kumar.
© 2022 American Geophysical Union. Published 2022 by John Wiley & Sons, Inc.
DOI: 10.1002/9781119481874.ch12

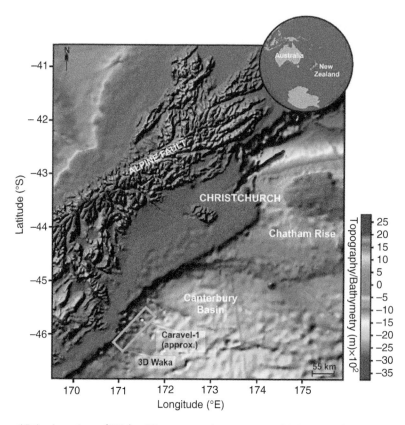

Figure 12.1 Location of Waka 3D prospect (green rectangle) in Canterbury Basin (CB), offshore New Zealand (NZ). The prospect is drilled by Caravel-1 well (after Bertoni et al., 2018); inset shows the global location.

surrounding sedimentary strata in the Waka prospect within the Canterbury Basin, offshore New Zealand. This chapter begins with a brief description of data followed by elaborating the results, derived from the computation of the SC and FIC meta-attributes approach as described in Chapters 9 and 10.

12.2. Description of the Data

High-resolution 3-D seismic data, acquired by New Zealand Petroleum Minerals (NZP&M), Govt. of NZ, are made available for academic research. The time-migrated seismic volume in the Waka prospect of Canterbury Basin off New Zealand comprises 800 inlines (NE–SW) and 5,755 crosslines (NW–SE) covering an area of 1,443 km^2. The data were acquired with 6 s record length, bin sizes of 25.0 m × 12.5 m, sampling rate of 4 ms (equivalent to Nyquist frequency of 125 Hz). It is apparent from the dominant seismic frequency of 25 Hz and an interval velocity of 5,500 m/s for the sill body (Hansen and Cartwright, 2006; Skogly, 1998) that the sills greater than 55 m in thickness can be resolved, as their top and base within the host-rock strata will generate discrete reflections. The presence of sill networks within the Waka prospect has been illustrated by Sahoo et al. (2015) and Reeves et al. (2018). The prospect is drilled by the well Carvel-1, which is restricted for public access. Approximate well location and geological tops are taken from the works of Sahoo et al. (2015) and Bertoni et al. (2018). The seismic data set is zero-phased, displayed in SEG American polarity convention. This means that a trough (red reflection) represents a downward decrease in acoustic impedance and a peak (black reflection) implies an increase in acoustic impedance.

12.3. Results and Interpretation

12.3.1. Data Enhancement, Attribute Analysis, and Neural Operation

The time-migrated 3-D seismic volume shows that the sill network and surrounding environment are accompanied by distorted seismic images (Figure 12.2a–b). Moreover, reflections are more chaotic within the plumbing zone. Conditioning the seismic data through DSMF enhances the lateral continuity of seismic events representing magmatic sill network, fluxed fluids, and other associated structures (Figure 12.2c–d). The geologic features within the conditioned volume are clearly visible (Figure 12.2c–d). Most of them are observed at shallow level within the Cretaceous to Intra-Eocene strata at depths of 2.4 to 3.1 s TWT (Figure 12.3a). The Paleocene through Intra-Eocene to Eocene package consists of layer bound polygonal faults which occur in different tiers. Moreover, submarine v-shaped canyons trending NE are developed at the seabed (Figure 12.3a). Magmatic sills within the Cretaceous to Intra-Eocene package exhibit saucer-shaped cross-sectional geometry with their limbs transgressing upwards and magma flux out from their tip into the overlying Eocene strata that has caused the forced folds (Figure 12.3b–e).

It is observed that sills are associated with high reflection strength, high texture contrast, and high entropy, as compared to the surrounding strata both in the western and eastern part of the prospect (Figure 12.4). Attribute analysis also suggests that the hydrothermal vent complex (HVC) and the sill-proximal areas are associated with low similarity and low frequency (Figure 12.5).

The neural training (Figure 12.6a–d) is quite satisfactory as evident from the minimum nRMS error and minimum misclassification % for both the train and test

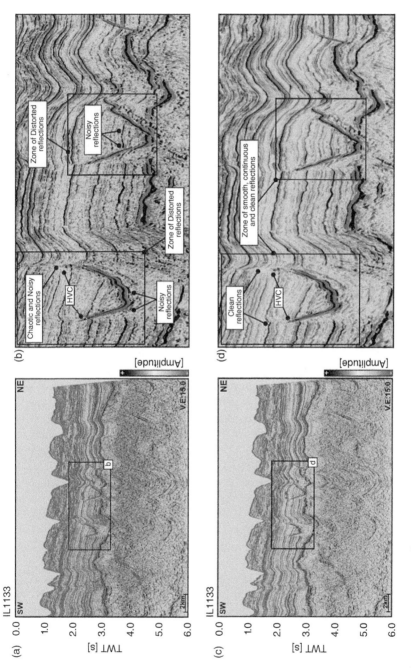

Figure 12.2 (a) Original time-migrated seismic section along inline (IL) 1133 from the Waka prospect. The magmatic sill complexes (rectangular box) are distorted. (b) Zoomed view of the rectangular box in (a) displaying obscured sill complexes and hydrothermal vent complex (HVC). (c) DSMF conditioned section demonstrating improvement of sill complexes and HVC (as marked by rectangular box in (a)). (d) Zoomed view of the rectangular box in (c) showing smooth, continuous, and distinct without any distortion.

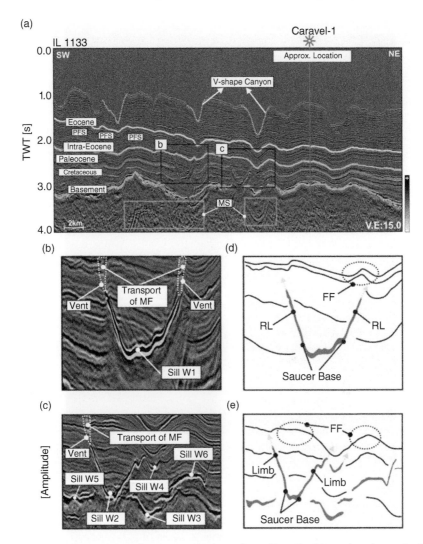

Figure 12.3 (a) Interpreted seismic section along line IL 1133 showing principal formations intersected by the wellbore Caravel-1. Reflections below the basement are poor and associated with migration smiles (blue rectangular box). (b and c) Zoomed versions of geometry and structural architecture of sills (for areas of rectangles b and c marked in (a)) are individually interpreted through line drawings (rectangles d and e). Sills exhibit saucer-shaped geometry with their limbs concave upwards. Transport of magmatic fluids through a locus of hydrothermal vent from sill tips into the overlying strata has resulted in forced fold structures. PFS: Polygonal Fault System; MS: Migration Smile; MF: Magmatic Fluid; RL: Rising Limb; FF: Forced Fold.

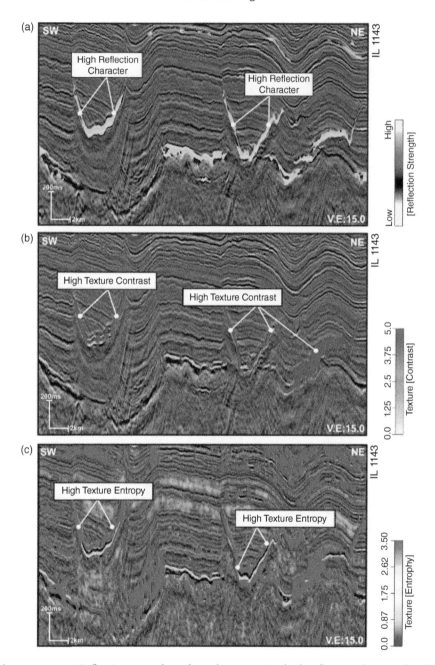

Figure 12.4   (a) Reflection strength attribute demonstrating high reflection character for sills as compared to surrounding sedimentary units. (b) Texture contrast attribute showing high contrast within the sill network. (c) Texture entropy attribute exhibiting higher entropy content within the sill complexes compared to the surrounding sedimentary environment.

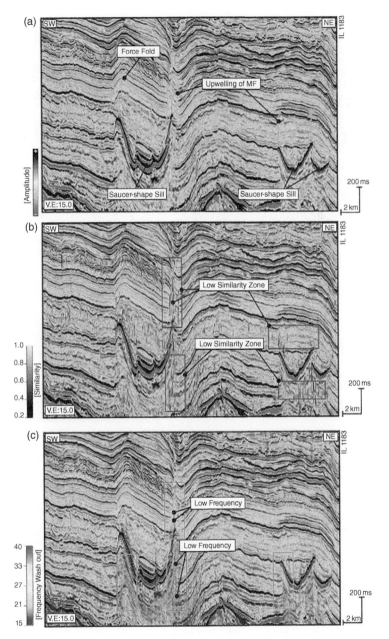

Figure 12.5   (a) Interpreted seismic section along line IL 1183 showing possible transport pathways (green arrows) of magmatic fluids into the overlying sedimentary succession. Hydrothermal vents are observed above the saucer-shaped sill terminations. (b) Similarity and (c) Frequency attributes exhibiting low similar and low frequency content (blue and green rectangles) within the fluid plumbing zones.

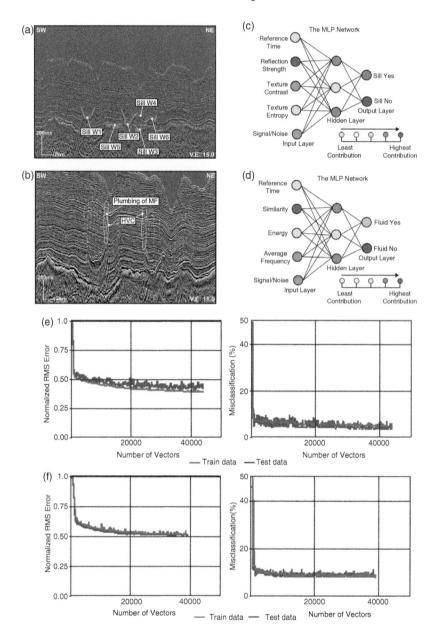

Figure 12.6 Example locations with (a) "sill-yes" and "sill-no," and (b) "fluid-yes" and "fluid-no." (c) and (d) MLP network consisting of 3 distinct layers: input with 5 nodes, hidden with 3 nodes, and output with 2, nodes respectively. Color variation (white through pale yellow and orange to red) within input layer shows relative contribution offered by the nodes during network operation. (e) and (f) nRMS error and misclassification (%) plots for the train and test data sets, respectively.

data sets for sills (Figure 12.6e) and fluid plumbing (Figure 12.6f) respectively. The nRMS error and misclassification % attains a minimum value of 0.45 and 0.5, and 5.87% and 8.16% for sills and a minimum value of 0.42 and 0.55, and 7.89% and 8.32% for fluid plumbing after 30 iterations. As usual, the output 0 means "sill-no" and "fluid-no," and 1 means "sill-yes" and "fluid-yes." It is important to achieve a minimum error for both the train and test data sets to make a closer approximation between the computed (that the system generates) and the targeted (that is based on the interpreter's knowledge on magmatic sill complexes and fluid plumbing system) data (Kumar et al., 2019a; Kumar & Sain, 2020). The relative contributions (i.e., weights) assigned to input nodes (i.e., seismic attributes) are shown in Tables 12.1 and 12.2.

12.3.2. Interpretation Through Sill Cube (SC) and Fluid Cube (FIC) Meta-Attributes

The extracted outcome, i.e., the SC meta-attribute, has brought out a saucer-shaped network of sills within 2.4 to 3.1 s TWT (Figure 12.7a) from seismic volume (Figure 12.7b). Magmatic sills (Sill W1, Sill W2, Sill W3, Sill W4 and Sill W6) exhibit concave upwards saucer-shaped cross-sectional geometry (Figures 12.7c–d, d–h, and f–j). However, the Sill W5 is strata concordant and

Table 12.1   Relative contributions offered by input nodes (seismic attributes) for neural training in designing the SC meta-attribute

Seismic attributes (input nodes)	Relative contribution (weights)
Reflection Strength	96.2
Signal/Noise	86.2
Texture Contrast	82.4
Reference Time	79.4
Texture Entropy	72.3

Table 12.2   Relative contributions offered by input nodes (seismic attributes) for neural training in designing the FIC meta-attribute

Seismic attributes (input nodes)	Relative contribution (weights)
Similarity	94.8
Average Frequency	85.8
Signal/Noise	84.2
Reference Time	77.5
Energy	73.8

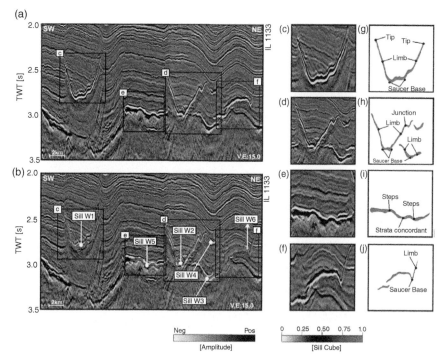

Figure 12.7 (a) Interpreted seismic section showing subsurface architecture of Waka prospect. Emplacement of sill complexes into the overlying sedimentary units results in forced folded strata. (b) SC meta-attribute co-rendered with interpreted seismic line. (c) and (d) Saucer-shaped sills exhibiting concave geometry with limbs rising upwards, and intersected to other sill through junction (as interpreted through line drawings in (g) and (h)). (e) Sills are connected through steps or bridges (as interpreted through line drawings in (i)). (f) Sill with saucer geometry having upward rising limbs (as interpreted through line drawings in (j)). Probability values of meta-attribute with a threshold limit of 0.75 are clipped over the section.

is connected through steps (Figure 12.7e–i). The saucer-shaped sills consist of transgressive inclined limbs through which magmatic fluids are fluxed out into the overlying sedimentary formations. Sill W5 represents an irregular morphology and occurs in between Sill W1 and Sill W2. Sills W1 and W6 lie towards the extreme western and eastern parts of the prospect, whereas the central part is dominated by Sills W2, W3, and W4. The sills identified through the SC meta-attribute cover a variable area of minimum ~1.08 km² for Sill W4, and maximum ~17.15 km² for Sill W2 within the Waka prospect (Table 12.3). Furthermore, the FIC meta-attribute qualifies its ability in demarcating the transport pathways of magmatic fluids through hydrothermal vents at the termination of sills (Figure 12.8a–b). The fluids vertically rise up to a height ~800 m. The emplacement of magmatic fluids is mostly observed in the central and eastern part. Several forced folds within the Eocene succession are observed, which tend to accommodate these emplaced fluids (Figure 12.8a–b). In the map view at t = 2.16 s

Table 12.3 Area of different sills delimited by a combined SC and FIC meta-attributes in the Waka prospect of Canterbury Basin off New Zealand

Sill name	Area covered (km^2)
Sill W1	12.07
Sill W2	17.15
Sill W3	3.84
Sill W4	1.08
Sill W5	2.76
Sill W6	1.52

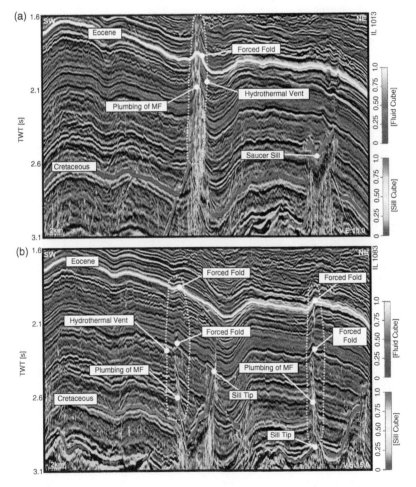

Figure 12.8 (a) and (b) Interpreted seismic sections along lines IL 1013 and IL 1083 showing subsurface magmatic emplacement caused by a sill network. Magmatic sills intrude through the Cretaceous units into the overlying Eocene sedimentary succession and flux out hot magma. Magmatic fluids are observed to rise vertically through hydrothermal vents causing forced folded structures. MF: Magmatic Fluid.

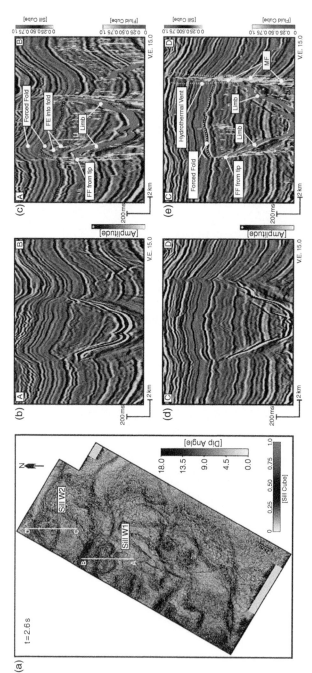

Figure 12.9 (a) Time slice at t = 2.60 s, displayed using co-rendered volumes of dip magnitude attribute and SC meta-attribute. The slice shows magmatic sill complexes namely Sill W1 and Sill W2 in the NW part. Their presence is confirmed from observations along random lines (AB and CD) prepared perpendicular to the trend of sills. (b) and (c) seismic sections through the random line AB exhibiting the presence of Sill W1 that is characterized by concave upward geometry with concordant base and discordant limbs. Vertical transport of magmatic fluids through the limbs has resulted in forced folded structures. (d) and (e) seismic sections through the random line CD showing the presence of Sill W2 characterized by concave upward geometry. Magmatic fluid vertically rises through the hydrothermal vent thereby doming the overlying strata. FF: Forced Fold; FE: Fluid Emplacement; MF: Magmatic Fluid.

(Figure 12.9a) the principal sills Sill W1 and W2 appear to trend ENE–WSW and ESE–WNW. Random lines AB and CD show their presence (Figure 12.9b–c) and reveal their structural geometry along with demarcation of fluid emplacements into the surrounding strata. This is clearly visible on the co-rendered images of SC and FIC meta-attributes with the amplitude data (Figure 12.9d–e). Such magmatic emplacements have resulted in doming up of the overlying strata and formation of forced folds that act as conducive structural traps (Figure 12.9d–e). Moreover, the 3-D view of SC meta-attribute (Figure 12.10) exhibit overall structural geometry of the magmatic sill (e.g., Sill W1) in the Waka prospect. The structure includes a concordant inner base with inclined sheets (or lobes) that spread out radially and flux magmatic fluids into the surrounding strata. The phenomena of magmatic fluid plumbing between the Cretaceous-Eocene strata is demonstrated through the 3D visualization using the meta-attributes in Figure 12.11.

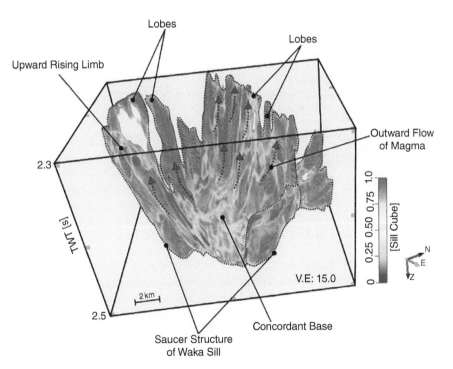

Figure 12.10   3-D view of the magmatic sill that has been delimited from 3D seismic volume by the SC meta-attribute approach. The sill shows a concordant inner base with a saucer-shaped geometry and upward rising limbs. The flanks contain inclined sheets or lobes rising upwards. Magmatic fluids are fluxed out through these lobes (indicated by blue dotted arrow) into the surrounding host strata. Probability value of meta-attribute with a threshold of 0.75 are displayed.

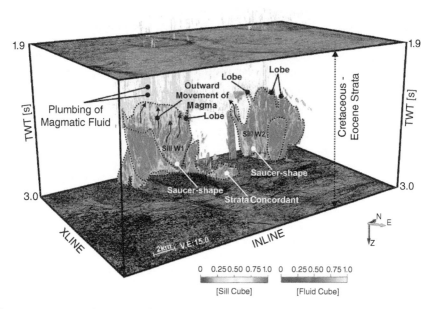

Figure 12.11   Emplacement of magma observed within the Cretaceous-Eocene strata at depths of 1.9–3.1 s TWT. Hot magma is transported through the Cretaceous units and further emplaced into overlying Eocene succession in the form of sill network. Magma is fluxed out from the saucer sills through lobes. The SC and FIC meta-attributes clearly bring out this geological scenario in the Waka prospect of Canterbury Basin. Probability value of meta-attribute is displayed with a threshold of 0.75.

12.3.3. Limitations of the Automated Approach

Meta-attribute interpretation is an automated approach that aims to improve the interpretation strategies by delimiting geologic targets from 3D seismic data based on neural networks (Kumar & Mandal, 2017; Kumar & Sain, 2018; Kumar et al., 2019a, 2019b; Singh et al., 2016). The disadvantage of this approach is that the data must be void of noises. Poor quality data may feed inaccurate example locations and hence lower the learning capability. Then the efficiency fails and the network generates artefacts, impeding the interpretation. This is true for the computation of any meta-attribute, and hence the interpretation of subsurface geologic features.

12.4. Summary

The SC and FlC meta-attributes have successfully captured the structural geometry of sill complexes and fluxed-out magmatic fluids within the Cretaceous to Eocene strata in Waka prospect of the petroliferous offshore Canterbury Basin. The main conclusions from this study are summarized as:

- Magma emplacement through the sill network has resulted in structural folds in overlying Eocene as well as younger geologic successions.
- Such magma routing has caused phreatic activity leading to circulation of superheated fluids that vertically rise to a height of ~800 m within the encased sedimentary strata.
- The approach is automated and can incorporate interpreters' intelligence in bringing out subsurface geometry of fluid plumbing from seismic data.
- Several new attributes may evolve in the future, which can be suitably included in refining the computation of meta-attribute for further advancement of interpretation.

References

Bertoni, C., Cartwright, J., Foschi, M., & Martin, J. (2018). Spectrum of gas migration phenomena across multilayered sealing sequences. *AAPG Bulletin*, *102*(6), 1011–1034. https://doi.org/10.1306/0810171622617210

Hansen, D. M., & Cartwright, J. (2006). Saucer-shaped sill with lobate morphology revealed by 3D seismic data: implications for resolving a shallow-level sill emplacement mechanism. *Journal of Geological Society*, *163*, 509–523. https://doi.org/10.1144/0016-764905-073

Kumar, P. C., & Mandal, A. (2017). Enhancement of fault interpretation using multi-attribute analysis and artificial neural network (ANN) approach: a case study from Taranaki Basin, New Zealand. *Exploration Geophysics*, *49*, 409–424. https://doi.org/10.1071/EG16072

Kumar, P. C., & Sain, K. (2018). Attribute amalgamation- aiding interpretation of faults from seismic data: An example from Waitara 3D prospect in Taranaki basin off New Zealand. *Journal of Applied Geophysics*, *159*, 52–68. https://doi.org/10.1016/j.jappgeo.2018.07.023

Kumar, P. C., Omosanya, K. O., & Sain, K., (2019a). Sill Cube: An automated approach for the interpretation of magmatic sill complex on reflection seismic data. *Marine and Petroleum Geology*, *100*, 60–84. https://doi.org/10.1016/j.marpetgeo.2018.10.054

Kumar, P. C., Sain, K., & Mandal, A. (2019b). Delineation of a buried volcanic system in Kora prospect off New Zealand using artificial neural networks. *Journal of Applied Geophysics*, *161*, 56–75. https://doi.org/10.1016/j.jappgeo.2018.12.008

Kumar, P. C., & Sain, K. (2020). Interpretation of magma transport through saucer sills in shallow sedimentary strata using an automated machine learning approach. *Tectonophysics*, *789*, 228541. https://doi.org/10.1016/j.tecto.2020.228541

Reeves, J., Magee, C., & Jackson, C. (2018). Unravelling intrusion-induced forced fold kinematics and ground deformation using 3D seismic reflection data. *Volcanica*, *1*(1), 1–17. https://doi.org/10.30909/vol.01.01.0117

Sahoo, T. R., Kroeger, K. F., Thrasher, G., Munday, S., Mingard, H., Cozens, N., & Hill, M. (2015). Facies Distribution and Impact on Petroleum Migration in the Canterbury Basin, New Zealand. Eastern Australian Basin Symposium, 187–202.

Skogly, O. P. (1998). Seismic characterisation and emplacement of intrusives in the Vøring Basin. M.Sc. Thesis, University of Oslo, Norway.

Singh, D., Kumar, P. C., & Sain, K. (2016). Interpretation of gas chimney from seismic data using artificial neural network: A study from Maari 3D prospect in the Taranaki basin, New Zealand. *Journal of Natural Gas Science and Engineering*, *36*, 339–357. https://doi.org/10.1016/j.jngse.2016.10.039

13

VOLCANIC SYSTEM INTERPRETATION

Volcanic systems are associated with igneous and sedimentary processes, which possess a two-sided effect on the petroleum system, e.g., maturity of source rocks; migration pathways for hydrocarbons; sealing and trapping mechanisms; geothermal history of the basin. As discovery of oil/gas becomes more difficult, oil industries look to difficult terrains like this for commercial gains. This chapter sheds light on the interpretation of a buried volcanic system in the Kora field of Taranaki basin, offshore New Zealand, by designing an Intrusion Cube meta-attribute from high resolution 3D seismic data. The meta-attribute distinctly reveals the structure of the buried volcano and several intrusive elements such as sill networks, dyke swarms, magmatic ascent, etc., in complex tectonic settings of the Kora field to enlighten on hydrocarbon exploration, geothermal story, and basin evolution.

13.1. Introduction

A volcanic system is defined as a set of interconnected geologic structures that confine a complete magmatic-sedimentary complex (Bischoff et al., 2017). The system is associated with igneous and sedimentary processes, which might have a two-sided effect on the petroleum system, e.g., maturity of source rocks; migration pathways for hydrocarbons; sealing and trapping mechanisms; geothermal history

Meta-Attributes and Artificial Networking: A New Tool for Seismic Interpretation,
Special Publications 76, First Edition. Kalachand Sain and Priyadarshi Chinmoy Kumar.
© 2022 American Geophysical Union. Published 2022 by John Wiley & Sons, Inc.
DOI: 10.1002/9781119481874.ch13

of the basin; etc. (Allis et al. 1995; Farrimond et al., 1999; Planke et al., 2005; Rateau et al., 2013; Rohrman 2007; Schutter, 2003; Stagpoole & Funnell, 2001; Sun et al., 2014; Zou, 2013). Such complexities may change the porosity and permeability of subsurface reservoirs and pose risks for exploration. The geological system must be assessed before taking a decision on exploitation of hydrocarbons. As discovery of oil/gas at ease is almost over, oil industries peer into difficult terrains like this for commercial gains. The interpretation of such a buried volcanic system is demonstrated by designing an Intrusion Cube (IC) meta-attribute from high-resolution 3-D seismic data in the Kora field of Taranaki Basin off New Zealand.

13.2. Research Workflow

The workflow is demonstrated in Figure 13.1, which consists of five different steps: (i) seismic data conditioning; (ii) selection of suitable seismic attributes; (iii) selection of example locations; (iv) designing a logical neural network for combining individual attributes into a single hybrid-attribute, called the Intrusion Cube (IC) meta-attribute; and (v) validating the outcome with well or other geoscientific data.

The seismic data is structurally conditioned using a pre-processed steering cube that stores the dip-azimuth information at each sample location. DSMF is applied to the data to improve the continuity of seismic events and suppress background random noises (Figure 13.2). These conditioned data are then used for the extraction of seismic attributes (Figure 13.3). A network is designed based on an ANN, output is validated, and the network is checked to see if it is trained properly before applying to the whole seismic volume.

At this stage, seismic attributes are grouped into three different cases (1, 2, and 3). The purpose of dividing the attributes into three different cases is to ascertain an optimum or best possible case that can capture the extension and distribution of plumbing system (consisting of sill networks, dyke swarms, ascent of magma and its emplacement) for realistic interpretation.

Case 1 consists of a set of attributes such as the coherency, energy, dip magnitude, azimuth, and curvature, which have been responsive in capturing igneous bodies from seismic data (Chopra & Marfurt, 2007; Pena et al., 2009; Zhang et al., 2011). Case 2, belonging to a set of attributes (a bit different from Case 1) , consists of energy, dip angle variance, similarity, envelope, and energy gradient, whereas Case 3 includes energy, dip angle variance, similarity, and envelope. The reference time and signal/noise (S/N) ratio are common to all cases. Attributes belonging to these three cases are parametrized at different step-outs and windows. Large, medium, and short vertical time windows (e.g., 160ms, 100ms, 80ms, 64ms, and 24ms) (Table 13.1) are used for designing the attribute such that the entire

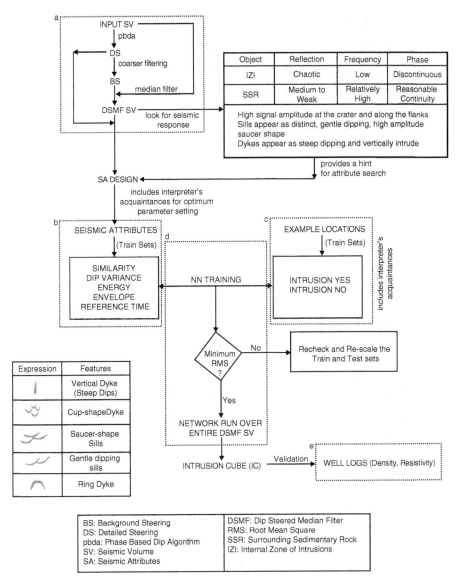

Figure 13.1   Workflow used for computation of the IC meta-attribute. Several expressions of intrusive features hint for interpretation (Modified after Mathieu et al., 2008). Apart from main attributes (mentioned in workflow), several other attributes (shown in Tables 13.3, 13.4 and 13.5) are also used.

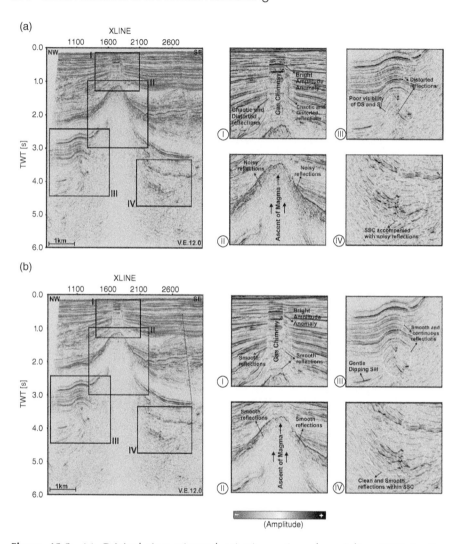

Figure 13.2 (a) Original time-migrated seismic section along inline 1317 in Kora prospect. Seismic signals within the plumbing system are distorted, as observed from four zoomed compartments (I, II, III, and IV), displaying them separately at immediate RHS; (b) DSMF time-migrated seismic section showing enhanced geologic features, as observed in four zoomed compartments (I, II, III, and IV) at immediate RHS.

geological responses (extension and distribution) of the targets are efficiently seized. The coherency mid-window, coherency short-window, energy mid-window, energy short-window, dip magnitude, azimuth, and most positive curvature attributes have been used in Case 1. On the other hand, the dip angle variance short-window, dip

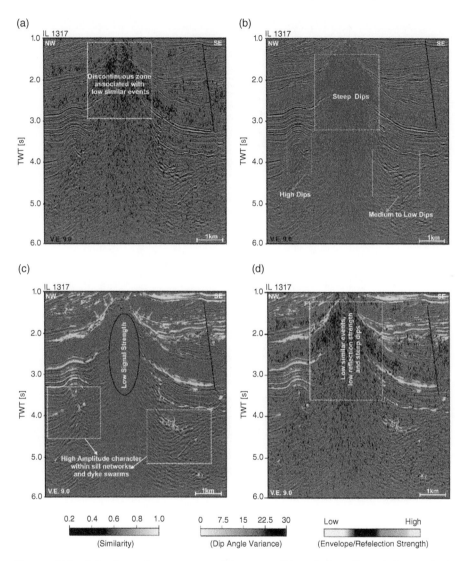

Figure 13.3 Seismic attributes extracted from conditioned data along inline 1317 in Figure 13.2. (a) Similarity attribute, co-rendered with amplitude data. (b) Dip angle variance attribute, co-rendered with amplitude data. (c) Energy attribute, co-rendered with amplitude data. (d) Co-rendered display of (a), (b), and (c) with amplitude data. The section is marked with a fault (black line).

Table 13.1 Key seismic attributes and their parameter settings for generating an optimized IC meta-attribute

Seismic attributes	Filter properties				Documented works
Similarity	DSMF and DS	[−32, +32] ms	(−1,0) & (1,0) 90° rotated	Output: Minimum Similarity	Tingdahl and de Rooij (2005); Kumar and Sain (2018)
Coherency	DSMF and DS	[−12, +12] ms; [−32, +32] ms	Cross Extension with 1×1 step-out	Output: Minimum Coherency	Chopra and Marfurt (2007) Kumar and Sain (2018)
Energy	DSMF	[−50, +50] ms; [−80, +80] ms; [−40, +40] ms	—	—	Chopra and Marfurt (2007)
Dip Angle	DS	—	—	—	Chopra and Marfurt (2007)
Azimuth	DS	—	—	—	Kumar and Sain (2018)
Dip Angle Variance	Dip Angle	[−12, +12] ms; [−32, +32] ms	Cube with 5×5 step-out	Output: Variance	Tingdahl and de Rooij (2005)
Envelope	DSMF	—	—	Output: Inst. Amp	Kumar et al., 2019a

angle variance long-window, energy short-window, energy mid-window, energy long-window, steered similarity, envelope, and energy gradient attributes are used in Case 2. Case 3 uses the dip angle variance short-window, dip angle variance long-window, energy short-window, energy mid-window, energy long-window, steered similarity, and envelope attributes.

The example locations are broadly grouped into "intrusion-yes" and "intrusion-no" classes (Figure 13.4a). The "intrusion-yes" classes are the set of all possible locations having intrusions, whereas the "intrusion-no" classes are all other locations devoid of any disturbance caused by intrusion. Around 1,650 example locations are picked up from a small volume of data. These locations are the same for all cases (Case 1, 2, and 3).

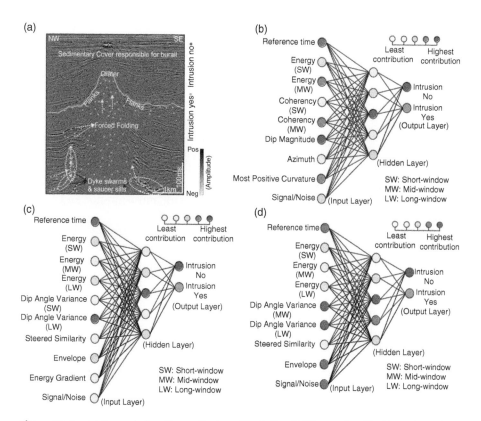

Figure 13.4   (a) Example locations, displayed for inline 1317, are classified into "intrusion-yes" (green dots) and "intrusion-no" (red dots) groups. The MLP networks designed for (b) Case 1, (c) Case 2, and (d) Case 3, consisting of three interconnected layers: input, hidden, and output layers.

Table 13.2   Neural network parameters used for three different IC meta-attribute models

	Neurons in different layers					
Neural model	Input layer	Hidden layer	Output layer	Learning rate	Momentum	Bias
IC Case 1	9	5	2	0.01	0.25	1
IC Case 2	10	5	2	0.01	0.25	1
IC Case 3	9	5	2	0.01	0.25	1

The neural network chosen for designing the meta-attribute is a fully connected MLP network (Aminzadeh & de Groot, 2006; Kumar & Sain, 2018; Meldahl et al., 2002). The perceptron is organized into different layers (Figure 13.4b–d). In the simplest form, the MLP consists of three distinct layers: input layer, hidden layer, and output layer. For Case 1, the input, hidden, and output layers consist of 9, 5, and 2 interconnected nodes. Case 2 has 10, 5, and 2 interconnected nodes. Case 3 consists of 9, 5, and 2 interconnected nodes. Other network parameters, e.g., learning rate and momentum, help in controlling the change required for designing the stable network for proper connection or link between the input, hidden, and output layers. Learning rate and momentum values are often assigned to the network by a trial-and-error basis (Poulton, 2001). The learning rate and momentum for all three cases are tuned to 0.01 and 0.25, respectively (Table 13.2). The sigmoid activation function takes input and squashes the output as 0s and 1s, where 0 refers to "intrusion-no" and 1 refers to "intrusion-yes."

The neural training begins by randomly selecting seismic lines (a small volume, may be 15–20% from the whole seismic volume) and splitting the selected data into train (70%) and test (30%) sets. Iterative neural training is performed to establish a minimum nRMS error between the neural output and desired output for the train data such that a probability output is obtained. The performance evaluation of the neural training is by calculating the nRMS errors and misclassification % (Figure 13.5). To check the performance of the network, the nRMS error and misclassification % are also looked into for the test data (Figure 13.5). Once satisfactory performance is achieved, the neural training is stopped and the output is quality checked over a few other randomly selected lines. After an acceptable outcome, the training is performed over the entire seismic volume to capture every single disturbance caused by the intruding body. Finally, this generates the IC meta-attributes for three different cases, the output of which is further investigated and validated with the available well data or other geological/geophysical information. Here the outcome is validated by using the density and resistivity logs of Kora-1A borehole that penetrated the volcanic formation as well as underlying lithological units.

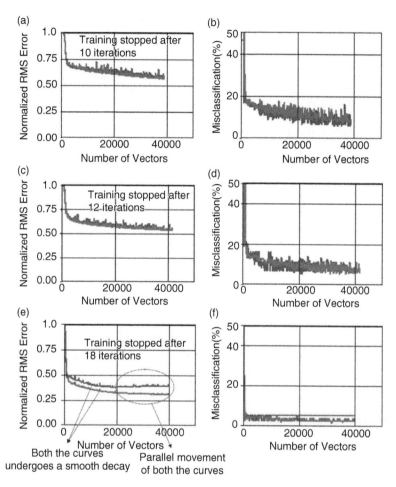

Figure 13.5 nRMS error and misclassification % for the train (red) and test (blue) data set (a, b) for Case 1, (c, d) for Case 2, and (e, f) for Case 3, respectively.

13.3. Results and Interpretation

13.3.1. Seismic Data Enhancement

It is observed that the buried volcano and other associated structural elements such as the sills and dykes are poorly visible in original data (Figure 13.2a). These features are masked with noisy events (I through IV compartments in Figure 13.2a). Structural conditioning of data through DSMF has reduced the effects of noises and enhanced lateral continuity of subsurface features

(Figure 13.2b). The seismic signal within volcanic gas chimneys, volcanic crater, and flanks, sill networks, and dyke swarms are improved (I through IV compartments in Figure 13.2b). Seismic attributes corresponding to these geological features have been efficiently captured. A few key seismic attributes (similarity, dip angle variance, and energy) that are very important for the interpretation of the buried volcanic system are demonstrated in Figure 13.3. This has been prepared by co-rendering extracted attributes with the amplitude data. The similarity attribute displays the discontinuous nature of seismic events within the zones at sub-volcanic depth and the Kora edifice (Figure 13.3a). The internal zone of edifice is characterized mostly by high/steep dips (Figure 13.3b) and associated with low energy (Figure 13.3c). The flanks of the edifice are associated with high-energy events (green arrows) due to large impedance contrast between the igneous materials and surrounding sedimentary rocks. The crest of the volcano shows high energy and amplitude events (green arrows). The responses captured by the similarity, dip angle variance, and energy attributes are more appreciated when co-rendered with the amplitude data (Figure 13.3d).

13.3.2. Neural Networks: Analysis and Optimization

The neural training shows the nRMS and misclassification % errors with iterations (Figure 13.5). The nRMS error attains a minimum value of 0.6 and 0.7 (Figure 13.5a) with a minimum misclassification % of 10.78% and 18.25 % (Figure 13.5b) after 10 iterations for Case 1. The relative contribution of input set (seismic attribute) are shown in Table 13.3. For Case 2, the nRMS error with a minimum value of 0.52 and 0.60 (Figure 13.5c), and minimum misclassification

Table 13.3 Sensitivity chart showing relative contribution offered by individual attribute to the neural system for Case 1 (LW: Long window; SW: Short window; MW: Mid window; MP: Most Positive)

Attributes	Weights
Dip magnitude	97.2
Coherency (MW)	94.8
Energy (MW)	92.1
Reference time	84.5
Curvature (MP)	81.9
Signal to noise (s/n)	78.5
Energy (SW)	62.5
Coherency (SW)	58.7
Azimuth (MW)	52.7

% of 6.21 and 11.03 (Figure 13.5d) are achieved after 12 iterations. The relative contribution of each input set (seismic attribute) for Case 2 is given in Table 13.4.

However, the nRMS error attains the minimum value of 0.3 and 0.4 (Figure 13.5e) with a minimum misclassification % of 2.38 and 5.13% (Figure 13.5f) after 18 iterations for Case 3. The training curves, compared to Case 1 and Case 2, exhibit a smooth decay pattern. The relative contribution provided by each input set for Case 3 is shown in Table 13.5. The extracted output for each case is now investigated by co-rending with amplitude to collude for an optimum meta-attribute for the best representation of the extension and distribution of the buried volcanic system. Different elements of the buried volcanic system, e.g., sill networks, dyke swarms, volcanic edifice, etc., are illustrated in Figure 13.6a.

Table 13.4 Sensitivity chart showing relative contribution offered by individual attribute to the neural system for Case 2 (LW: Long window; SW: Short window; MW: Mid window; MP: Most positive)

Attributes	Weights
Dip angle variance (SW)	98.5
Reference time	83.5
Signal to noise (s/n)	78.5
Steered similarity	77.8
Dip angle variance (LW)	75.0
Envelope	73.2
Energy (SW)	72.1
Energy (LW)	68.2
Energy (MW)	58.5
Energy Gradient	42.4

Table 13.5 Sensitivity chart showing relative contribution offered by individual attribute to the neural system for Case 3 (LW: Long window; SW: Short window; MW: Mid window; MP: Most positive)

Attributes	Weights
Dip angle variance (SW)	99.2
Reference time	89.5
Envelope	82.4
Signal to noise (s/n)	80.3
Steered similarity	74.5
Energy (SW)	73.9
Dip angle variance (LW)	70.5
Energy (LW)	68.2
Energy (MW)	62.5

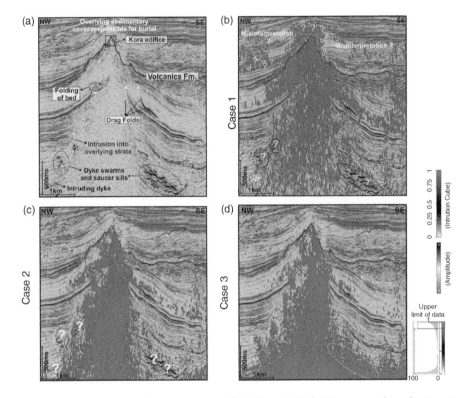

Figure 13.6 (a) Interpreted seismic section for inline 1240. (b) IC meta-attribute for Case 1 shows image of the buried volcanic system at the cost of some artefacts or misinterpretation outputs. (c) IC meta-attribute for Case 2 presents the image of the buried volcanic complex, though devoid of pitfalls but fail to capture dyke swarms and sill networks. (d) IC meta-attribute for Case 3 brings out an optimal view of the buried volcanic system along with capturing the geometry of magmatic emplacement bodies.

Emplacements of these structural elements into the host sedimentary units modulate the subsurface structural architecture, resulting in jacking up sedimentary strata, drag folds, etc. The top of the volcanic formation is marked with a black solid line. It is observed that the meta-attribute for Case 1 shows several artefacts (green dotted oval with yellow question mark in Figure 13.6b) that are against the regional geology. Most of the portions above the top of the volcanic interval are linked with high IC values, suggesting further extension of the buried volcano. By comparing with regional geology (Figure 13.6a), it is observed that the sequence above the volcanic interval is a thick sedimentary cover of the Giant Foresets that has buried the volcanic complex. Towards the deeper part of the section, we see that the IC meta-attribute for Case 1 fails to capture the intruding elements, e.g., dyke swarms and sills (yellow and blue dotted oval with white question mark in Figure 13.6b).

The meta-attribute for Case 2 (Figure 13.6c) efficiently images the buried subsurface system and honors the regional geology of the study area. The shallower parts representing thick sedimentary cover are devoid of artefacts. However, the IC meta-attribute fails to capture the geometrical shape and orientation of dyke swarms and sills in the deeper section (yellow and blue dotted ovals with white question mark in Figure 13.6c). The meta-attribute for Case 3 (Figure 13.6d) enlightens the extension and distribution of the buried volcanic system along with other structural elements such as the dyke swarms and sills. The IC meta-attribute volume based on Case 3 has brought out an optimum and reliable picture of the subsurface buried volcanic system from 3D time-migrated seismic data.

13.3.3. Geologic Interpretation Using the IC Meta-Attribute

An attempt has been made to interpret the buried volcanic complex consisting of a complex plumbing system that includes saucer shape sills, dyke swarms, and magma chambers with the IC meta-attribute approach. The structural elements are mostly observed in the deeper part of the seismic section (2.0–5.0 s TWT) (Figures 13.7 and 13.8). The sills in the NW and SE parts demonstrate saucer-shaped patterns (Figure 13.8a, c, and e) and exhibit concave upwards cross-sectional architecture. Moreover, sills in the SE part exhibit broken bridge and stepped geometries. The vertically intruding dyke swarms restrict their appearance to the NW part only. It is observed that the sills in isolation (NW part in Figure 13.8a) cause little deformation to the overlying sedimentary strata. The host sedimentary units are intruded by sill networks and dyke swarms (white dotted oval in the deeper part). The flanks of the Kora edifice are faulted and several radial faults are observed to diverge out from the center of the edifice. Ascent of magma (AOM) (white dotted arrow) is observed at sub-volcanic depth of the Kora edifice. The change in structural architecture of the overlying sedimentary formations is clearly observed in the NW part (Figure 13.8c). Movements of intrusive elements (dyke swarms and saucer sills, as indicated by yellow and white dotted ovals) into the host sedimentary units are shown by yellow arrows. Drag folds (green dotted oval and white dotted arrows) that have resulted from magmatic emplacement into the sedimentary strata are also observed. The complex plumbing system is observed beneath the volcanic edifice (Figure 13.8a, c, and e), which possesses discontinuous character and crosscuts the sedimentary unit below the volcanic interval.

The magmatic emplacement into the host sedimentary interval (or the pre-magmatic sequence) causes large structural deformation, and most of these activities are overserved in the sub-volcanic depth of 2.5–3.5 s TWT (Figures 13.7 and 13.8). Emplacement to the sedimentary formations has resulted in the development of forced folds exhibiting double folded geometry (Figure 13.8e). These structures are tilted away from the primary vent of the Kora edifice. The IC meta-attribute (Figure 13.8b, d, and f) has captured and brought out these.

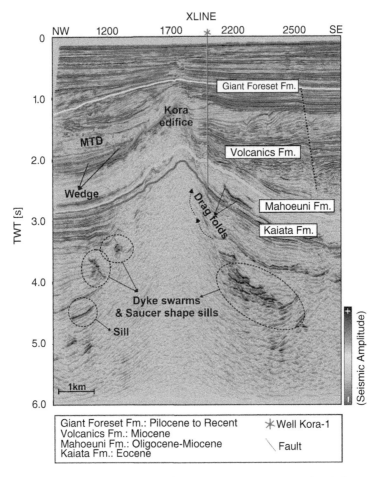

Figure 13.7   Interpreted seismic section for inline 1240 superimposed with the Kora-I well (solid red line with star at top) that has penetrated the volcanic formation through flanks of the Kora edifice. The tops of different geologic formations are marked by different colors. Intrusions into the host sedimentary units have brought out changes in the subsurface structural architecture (causing jacking up of sedimentary strata) (V.E:9.0).

The geometrical architecture of saucer-shaped sills and vertically intruding dyke swarms are clearly imaged by the IC meta-attribute. These structures are associated with high IC values. Interpreted magmatic plumbing system and pathways of magmatic emplacement within the sedimentary intervals are prominently demarcated by the IC meta-attribute (Figure 13.8b, d, and f). The Kora edifice, associated with high values of the IC meta-attribute, reveals the vertical and lateral extension of the volcano.

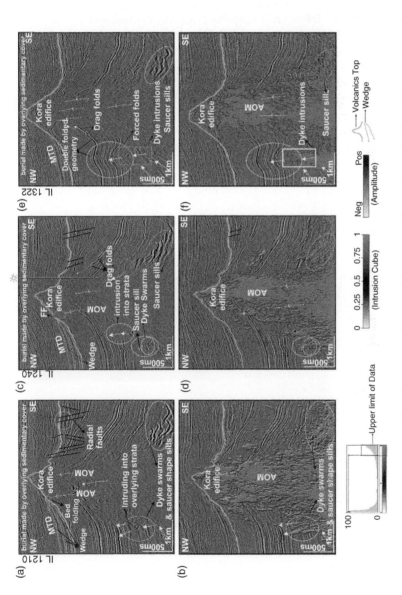

Figure 13.8 (a) Interpreted seismic section for inline 1210. The top of the Kora edifice is shown by the solid cream line. (b) IC meta-attribute, co-rendered with amplitude data for the same line, distinctly showing the activities. (c) Interpreted seismic section for inline 1240. (d) IC meta-attribute, co-rendered with amplitude data for the same line, clearly illuminating the activities. (e) Interpreted seismic section for inline 1322. (f) IC meta-attribute, co-rendered with amplitude data for the same line, evidently showing the extension and distribution of the @ buried volcanic system.

Dyke swarms and sills act as prominent structures for emplacing magma into the overlying sedimentary units. This has resulted in forced folds, double folded geometry pattern, and drag folds observed in the NW (white dotted ovals and black arrows) part of the study area. These observations are illustrated by the yellow rectangular box and white dotted oval. The movements of magmatic fluids are indicated by yellow arrows (Figure 13.8a–f). The IC meta-attribute has captured these events within the study area. This is better understood from a time slice at t = 2.7 s, viewed with the energy gradient attribute (Figure 13.9a). The IC meta-attribute, co-rendered with the energy gradient attribute, demonstrates high IC values within the Kora edifice and nearby forced folds (Figure 13.9b). A random line AB confirms these observations, where dykes are perceived to intrude into the overlying sedimentary units (green dotted arrow and white dotted oval), causing double folded geometry of geological beds (yellow dotted oval in Figure 13.9c).

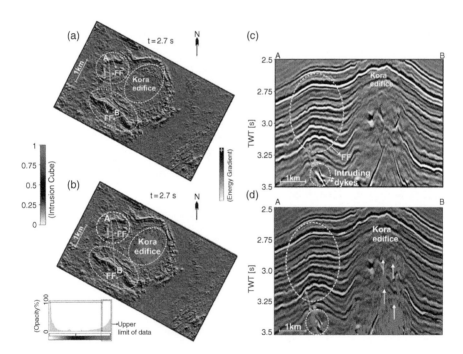

Figure 13.9 (a) Time slice at t = 2.7 s, displayed with energy gradient attribute, showing the structural architecture of the Kora edifice (white dotted oval) and associated domed strata like the forced folds (FF) (yellow dotted oval). (b) Energy gradient attribute, co-rendered with IC meta-attribute for the same time slice. High IC values are observed within the Kora edifice. (c) Seismic section through the random line AB, as marked in (a–b). (d) IC meta-attribute clipped over the seismic section along the same line confirms the edifice and validates the interpretation.

The IC meta-attribute, co-rendered with amplitude data, highlights all these facts (Figure 13.9d). Furthermore, a time slice at t = 4.2 s demonstrates the structural geometry of sill complex and dyke swarms (Figure 13.10a). The energy gradient attribute, co-rendered with the IC meta-attribute for the same time slice, shows high IC values within the volcanic core, sill complex, dykes, and the surrounding areas (Figure 13.10b). The sill complex and dyke swarms are distributed in the eastern and western parts of the buried volcanic complex. Random lines CD and EF illustrate the nature and geometry of subsurface structures (Figure 13.10c–d).

Magmatic ascent into shallow sedimentary intervals on the flanks of the Kora edifice results in a forced fold dome. This is clearly observed from the co-rendered plot of dip angle, azimuth, and IC attributes over time slice t = 2.19 s TWT (Figure 13.11a). This forced dome structure is linked with several radial faults (black arrows) that diverge out from the volcanic core (white dotted circle). Forced folds (FF) (green arrow) are observed closer to the flanks of the Kora edifice. Random lines GH, IJ and KL confirm these interpretations (Figure 13.11b–d).

The top of main vent complex (MVC) and the flanks are associated with high amplitude, and are characterized with an amalgamation of coherent and disrupted reflections that diverge out radially from the center (Figure 13.7 and Figure 13.12a and b). The volcanic core is associated with high IC values (Figure 13.12a). The lava flows are observed to diverge along the flanks from the center of the volcanic edifice. The IC-meta attribute illuminates the MVC of the Kora edifice, visualized both in plan (Figure 13.8) and map view (Figure 13.12a). The random line MN drawn perpendicular to the volcanic center illuminates the Kora edifice (Figure 13.12b). The deposits of lava flow mimic the channelized geometry pattern and diverge out radially from the center of eruptive vent (Figure 13.12a and c). It is observed that the thickness of these deposits varies between 34.92 m and 44.63 m for an assumed interval velocity of 5,500 m/s (Skogly, 1998).

13.3.4. Validation of the IC Meta-Attribute

The inferences drawn using the IC meta-attribute are validated through the log signatures of the Kora-1 exploratory well, which was drilled within the study region by ARCO Petroleum. The well was drilled up to a depth of 3,241 m to verify the presence of Eocene Tangora sandstone units within a large-dome structure associated with the Miocene volcanics. It is reported that hydrocarbons were encountered at the top of the Miocene volcanics and upper and lower Tangora sandstone units. A production test within the Miocene volcanic section delivered 1,168 barrels of oil per day (bopd) (ARCO Petroleum, 1988). The density and resistivity logs (Figure 13.13a) are used to validate the IC meta-attribute computed along a seismic line (Figure 13.13b) passing through the borehole. The borehole penetrated through the Miocene (Volcanics formation; 1,781.2 m), the Oligocene (TeKuiti formation; 2,898.5 m), and the Oligocene-Miocene

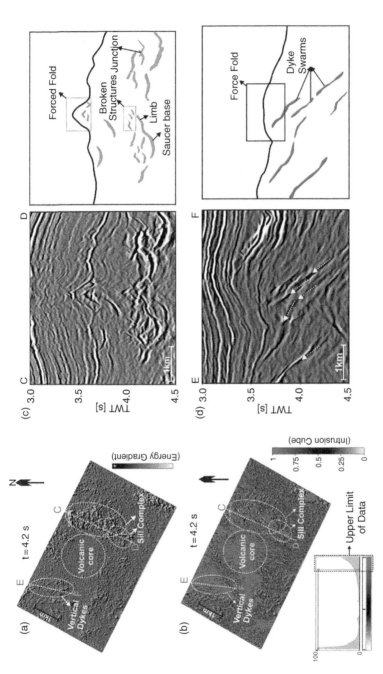

Figure 13.10 (a) Time slice at t = 4.2 s, displayed with the energy gradient attribute, showing the structural architecture of saucer-shaped sill complex and dyke swarms (white dotted ovals). The volcanic core is indicated by the white dotted circle. (b) Energy gradient attribute, co-rendered with IC meta-attribute for the same time slice. (c) Seismic section along the random line CD, as marked in (a–b), which highlights the relevance of sill networks. The sills exhibit a saucer shape and broken bridge patterns; an interpreted sketch is shown immediately to the right. (d) Seismic section along another random line EF, as marked in (a–b), which highlights the presence of vertically intruding dykes (marked with white dotted arrows; an interpreted sketch is shown immediately to the right).

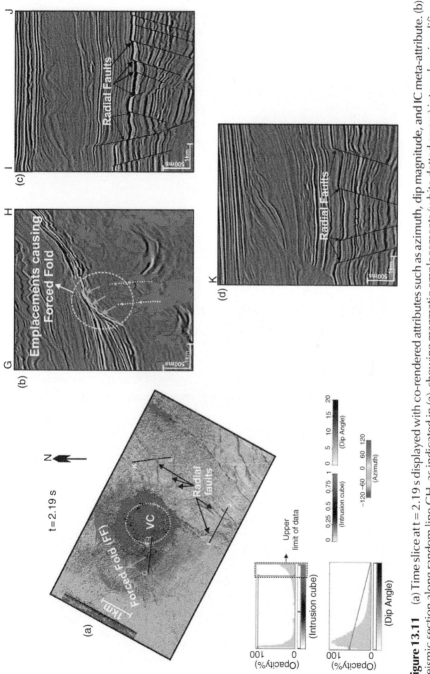

Figure 13.11 (a) Time slice at t = 2.19 s displayed with co-rendered attributes such as azimuth, dip magnitude, and IC meta-attribute. (b) Seismic section along random line GH, as indicated in (a), showing magmatic emplacements (white dotted arrow) into volcanic edifice, resulting in the doming of flanks creating forced folds (FF). The domed portion is indicated with the solid cream line; seismic sections along the random lines (c) IJ and (d) KL, as indicated in (a), show radial faults diverging out from the volcanic core (VC).

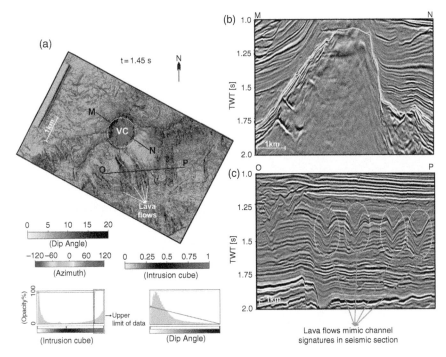

Figure 13.12   (a) Time slice at t = 1.45s displayed with co-rendered attributes (azimuth, dip magnitude, and IC meta-attribute). (b) Seismic section along random line MN passing through volcanic core (VC), as marked in (a). (c) Seismic section along another random line OP passing through away from the VC confirms the interpretation made from (a).

(Mahoeuni formation; 2,581.5 m), and terminated at the end of the Eocene (Kaiata formation; 3,241.8 m).

The zone lying between ~1,781 m and 2,375 m depth was prone to severe igneous activities throughout the entire Miocene period, as observed by high density and high resistivity values in the log data. The superposition of logs over the seismic section indicates this phenomenon. The depth interval between ~2,375 m and 2,750 m, associated with low density and low resistivity, suggest that though magmatic activity took place, it was not so intense during the Eocene-Oligocene period. However, relics of these activities comprising the plumbing system, magmatic feeder, AOM, etc., prevailed within these periods, which are exhibited by the IC meta-attribute. The zones between ~2,750 m to 3,241 m, accompanied by high density and high resistivity, and further below are associated with high IC meta-attribute. Such a magmatic event that took place over a large geological period from the Late Cretaceous through Paleocene, Eocene, and Oligocene to Miocene had not only modulated the architecture of the surrounding environment but also generated several structural traps conducive for engulfing hydrocarbon resources.

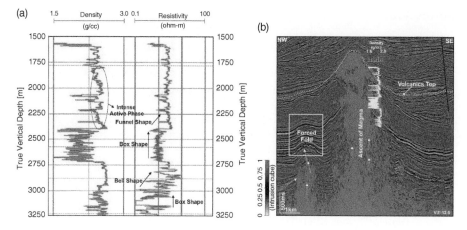

Figure 13.13   (a) Density log (red) and resistivity log (pink) in the Kora-1 borehole drilled within the study region. Different stratigraphic horizons are marked (violet dotted line for Kaiata Fm.; pink dotted line for TeKuiti Fm.; green dotted line for Mahoenui Fm.; orange dotted line for Volcanic Fm.). Volcanic formation that remained active during the Miocene period is associated with high density and high resistivity. (b) Seismic section, co-rendered with the IC meta-attribute and superimposed with density log data, showing different elements along with the AOM. (Fm.: Formation)

13.4. Summary

The main conclusions are summarized as:

- An attempt has been made to consistently improve the interpretation strategies such that misinterpretations and pitfalls can be minimized and a meaningful geological interpretation can be made.
- The IC meta-attribute obtained from Case 3 is an optimum approach to such an objective, where the geological targets such as buried volcanoes are distinctly illuminated from 3-D seismic volume.
- The IC meta-attribute has provided a realistic interpretation of subsurface geology from time-migrated seismic data in the Taranaki Basin off NZ by delimiting the extension and distribution of subsurface plumbing system.
- The study has brought out a buried volcano and several intrusive elements such as sill networks, dyke swarms, magmatic ascent, etc., in complex tectonic settings of the Kora field.
- The IC meta-attribute can be generated in any global environment from an optimal set of seismic attributes (as showcased in this study). It may depend on the geologic objects to be interpreted and the tectonic regime that surrounds them.

References

Allis, R. G., Armstrong, P. A., & Funnell, R. H. (1995). Implications of high heat flow anomaly around New Plymouth, North Island, New Zealand. *New Zealand Journal of Geology and Geophysics, 38*, 121–130. https://doi.org/10.1080/00288306.1995.9514644

Aminzadeh, F., & De Groot, P. (2006). *Neural networks and other soft computing techniques with applications in the oil industry*. EAGE Publications.

ARCO Petroleum, (1988). Final well report, Kora-1 and Kora-1A, PPL 38447, Ministry of Economic Development New Zealand. Unpublished Petroleum Report Series, PR 1374, 1–885.

Bischoff, A. P., Andrew, N., & Beggs, M. (2017). Stratigraphy of architectural elements in a buried volcanic system and implications for hydrocarbon exploration. *Interpretation, 5*, SK141–159. https://doi.org/10.1190/INT-2016-0201.1

Chopra, S., & Marfurt, K. J. (2007). *Seismic attributes for prospect identification and reservoir characterization*. SEG, Tulsa.

Farrimond, P., Bevan, J. C., & Bishop, A. N. (1999). Tricyclic terpane maturity parameters: response to heating by an igneous intrusion. *Organic Geochemistry, 30*, 1011–1019. https://doi.org/10.1016/S0146-6380(99)00091-1

Kumar, P. C., & Sain, K. (2018). Attribute amalgamation-aiding interpretation of faults from seismic data: An example from Waitara 3D prospect in Taranaki basin off New Zealand. *Journal of Applied Geophysics, 159*, 52–68. https://doi.org/10.1016/j.jappgeo.2018.07.023

Kumar, P. C., Omosanya, K. O., & Sain, K. (2019). Sill Cube: An automated approach for the interpretation of magmatic sill complexes on seismic reflection data. *Journal of Marine and Petroleum Geology, 100*, 60–84. https://doi.org/10.1016/j.marpetgeo.2018.10.054

Mathieu, L., De Vries, B. V. W., Holohan, E. P., & Troll, V. R. (2008). Dykes, cups, saucers and sills: Analogue experiments on magma intrusion into brittle rocks. *Earth and Planetary Science Letters, 271*, 1–13. https://doi.org/10.1016/j.epsl.2008.02.020

Meldahl, P., Najjar, N., Oldenziel-Dijkstra, T., & Ligtenberg, H. (2002). *Semi-automated detection of 4D anomalies*. Paper presented in 64th EAGE Conference & Exhibition (pp. cp-5). European Association of Geoscientists & Engineers. https://doi.org/10.3997/2214-4609-pdb.5.P315

Pena, V., Chávez-Pérez, S., Vázquez-García, M., & Marfurt, K. J. (2009). Impact of shallow volcanics on seismic data quality in Chicontepec Basin, Mexico. *Leading Edge, 28*, 674–679. https://doi.org/10.1190/1.3148407

Planke, S., Rasmussen, T., Rey, S. S., & Myklebust, R. (2005). Seismic characteristics and distribution of volcanic intrusions and hydrothermal vent complexes in the Vøring and Møre basins. In: *Petroleum Geology Conference series* (Vol. 6, pp. 833–844). Geological Society, London. https://doi.org/10.1144/0060833

Poulton, M. M. (Ed.). (2001). *Computational neural networks for geophysical data processing*. Elsevier.

Rateau, R., Schofield, N., & Smith, M. (2013). The potential role of igneous intrusions on hydrocarbon migration, West Shetland. *Petroleum Geoscience, 19*, 259–272. https://doi.org/10.1144/petgeo2012-035

Rohrman, M. (2007). Prospectivity of volcanic basins: Trap delineation and acreage de-risking: *AAPG Bulletin*, *91*, 915–939. https://doi.org/10.1306/12150606017

Schutter, S. R. (2003). Hydrocarbon occurrence and exploration in and around igneous rocks: *Geological Society of London, Special Publications*, *7–33*. https://doi.org/10.1144/GSL.SP.2003.214.01.02

Skogly, O. (1998). *Seismic characterization and emplacement of intrusives in the Vøring Basin. Cand Scient thesis, Department of Geology*, University of Oslo.

Stagpoole, V., & Funnell, R. (2001). Arc magmatism and hydrocarbon generation in the northern Taranaki Basin, *New Zealand, Petroleum Geoscience*, *7*, 255–267. https://doi.org/10.1144/petgeo.7.3.255

Sun, Q., Wu, S., Cartwright, J., Wang, S., Lu, Y., Chen, D., & Dong, D. (2014), Neogene igneous intrusions in the northern South China Sea: Evidence from high-resolution three-dimensional seismic data. *Marine and Petroleum Geology*, *54*, 83–95. https://doi.org/10.1016/j.marpetgeo.2014.02.014

Tingdahl, K. M., & de Rooij, M. (2005). Semi-automatic detection of faults in 3D seismic data. *Geophysical Prospecting*, *53*, 533–542. https://doi.org/10.1111/j.1365-2478.2005.00489.x

Zhang, K., Marfurt, K. J., Wan, Z., & Zhan, S. (2011). Seismic attribute illumination of an igneous reservoir in China, *Leading Edge*, *30*, 266–270. https://doi.org/10.1190/1.3567256

Zou, C. (2013). *Volcanic reservoirs in petroleum exploration*. 1st edn. Elsevier. https://doi.org/10.1016/C2011-0-06248-8

14

INTERPRETATION OF MASS TRANSPORT DEPOSITS

Submarine mass movement generates different depositional units that include creep, slides, slumps, debris flows, etc., which can be collectively termed the mass transport complexes or Mass Transport Deposits (MTDs). Their interpretation is crucial, as such deposits during translation over the instable slope may lead to several catastrophic submarine events, e.g., landslides, tsunamis, or avalanches, and hence possess precursory threats for subsea installations. This chapter shows how to design a workflow and to compute a new attribute, the MTD cube meta-attribute, through an artificial neural network. This has illuminated the structural architecture of the MTD from 3-D seismic data in the Karewa prospect of offshore Taranaki Basin, New Zealand.

14.1. Introduction

Slope failure or slope instability results in the transportation of unconsolidated sediments into the deep water environment under the influence of gravitational force (Omosanya, 2018; Posamentier & Kolla, 2003; Varnes, 1978). During the failure of a submarine slope, material (or mass) flows from the regions of high gradient, e.g., shelf-slope break and upper slope, and gets translated downslope above a surface called the basal shear surface (BSS) (Bull et al., 2009; Martinsen, 1994; Varnes, 1978) through a number of mass movement processes. The BSS within these deposits gets developed due to progressive shear failure during

Meta-Attributes and Artificial Networking: A New Tool for Seismic Interpretation,
Special Publications 76, First Edition. Kalachand Sain and Priyadarshi Chinmoy Kumar.
© 2022 American Geophysical Union. Published 2022 by John Wiley & Sons, Inc.
DOI: 10.1002/9781119481874.ch14

sediment translation. Such submarine mass movement generates different depositional units that include creep, slides, slumps, and debris flows, which can be collectively termed the mass transport complexes or Mass Transport Deposits (MTDs). Their interpretation is crucial, as such deposits during translation over the instable slope may lead to several catastrophic submarine events, e.g., landslides, tsunamis, and avalanches, and thus possess precursory threats for subsea installations.

It is necessary to design a new attribute, termed the MTD cube (MTDC) meta-attribute, which can efficiently illuminate the subsurface structural geometry of the MTD from reflection seismic data. It starts by presenting a workflow suited to the computation of this meta-attribute from a set of other seismic attributes. The seismic data from the Karewa field, offshore Taranaki Basin, has been used to demonstrate the efficacy of the MTDC meta-attribute for advanced interpretation.

14.2. Data and Research Workflow

The data used for this research includes time-migrated 3-D seismic data that consists of 393 inlines (Line no. 1,000 to 1,393) and 2,000 xlines (Line no. 2,800 to 4,800). The seismic data, which was acquired by PGS M/V Orient Explorer, covers an area of ~ 122 km². Additional acquisition parameters include bin spacing of 25.0 m × 12.5 m (inl/xrl), 4 ms sampling interval, and 5 s record length. Closer xline spacing is used to capture more structural details of the target from the dip direction. The data are displayed using SEG American polarity convention where an increase in acoustic impedance is represented by a peak (positive amplitude black on seismic sections). For a dominant frequency of 40 Hz within the Karewa MTD and assuming sediment velocity of 1,800 m/s, the limit of vertical resolution ($\lambda/4$) is calculated as 10 m.

The step-by-step approach (Figure 14.1) consists of: (i) structural enhancement of 3D seismic volume; (ii) computation of suitable seismic attributes; (iii) selection of training locations; (iv) setting up a logical neural network for the computation of MTDC meta-attribute; and (v) validation of MTD outcome with published literature and available petroleum report of well data.

The seismic data is optimally conditioned using a structure-oriented filter (SOF) that utilizes a pre-computed dip-azimuth volume to steer the data in the direction of local dip of the seismic events (Tingdahl, 1999). Then a DSMF is applied to filter the data. This is performed using a 3×3 median filtering step-out. This conditioned data along with the dip-azimuth steering volume are then used for extraction of attributes followed by designing an ANN architecture and validating the outcome. A suite of seismic attributes such as similarity, dip

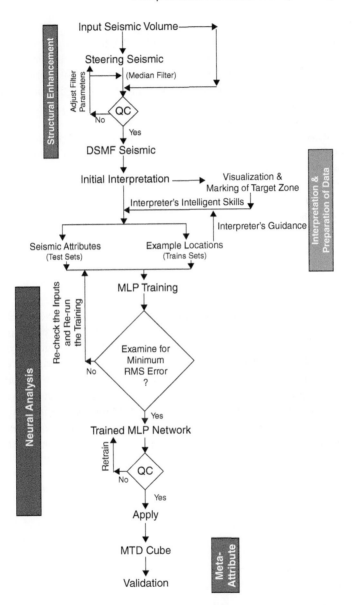

Figure 14.1   Workflow used for the computation of the MTDC Meta-attribute.

angle variance, energy, average frequency variance attributes, etc., are selected to capture the characteristics of MTD from the seismic volume. It is crucial to parametrize the selected attributes so that they are able to arrest the responses of the MTD target. The seismic attributes are defined into the neural system using three vertical time windows (large: 80 ms, medium: 32 ms, and short: 24 ms) and 6 by 6 inline/xline step-outs (i.e., 6 traces in inline and 6 traces in xline directions) for efficient representation of the target.

The training/example locations are selected randomly along a few xlines and inlines, and the "MTD-yes" and "MTD-no" targets/objects are defined based on seismic properties and geologic characteristics, as described earlier. The MTD targets are assigned with "1" and the non-MTD targets are associated with "0" according to the binary classification rule. Around 755 objects and 745 non-objects are picked for training and testing from the selected lines, and are assigned binary numbers 1 and 0 respectively.

A fully connected MLP network is designed for the computation of a hybrid attribute from a set of selected attributes that are related to the MTD. The hybrid attribute is defined as the MTDC meta-attribute, The MLP for this work consists of three distinct layers: the input, the hidden, and the output layers. The designed MLP network contains 5 neurons in input layer, 3 in hidden layer, and 2 in output layer, which are interconnected. Other network parameters such as the learning rate and momentum are optimally set to 0.01 and 0.25, respectively, through several trials. The activation function used in this study is also a sigmoid function, as has been used earlier. Only 70% of the picked data (which is again only a randomly selected small volume (15-20%) of entire volume of data) are used for training in which the related seismic attributes are taken as input to compute the response lying between 0 and 1 using a feedforward process. The network parameters (rate of learning, momentum, and most importantly the weights) are automatically adjusted iteratively based on a backpropagation algorithm to minimize the difference between the network's response and desired output for the train data. Since the process computes responses at the remaining 30% of locations (test data) also, the difference between the prediction and desired output is also calculated simultaneously for the test data to see if the model is predicting properly by observing the nature of difference curve, i.e. the decreasing trend of differences with iterations. Iterative neural training is continued until a minimum nRMS error and misclassification % is achieved between the observed and computed data such that a probability output is obtained at all chosen locations (Kumar & Sain, 2018; Kumar et al., 2019a; 2019b; Kumar & Sain, 2020). The performance of the network is quality checked through visual inspection by clipping the computed meta-attribute over other seismic lines. Once satisfied with the quality performance, the network is made to run over the entire seismic cube such that human intervention is limited and the process of interpretation is accelerated. This results in the MTD cube meta-attribute, and automatically delineates the distribution and geometry of the MTD in 3-D space.

14.3. Results and Interpretation

We mainly focus on the interpretation of seismic reflection data between 1.25 to 2.0 s TWT (Figure 14.2), within which lies the Karewa slump zone, overlain by the Plio-Pleistocene sequence (PPS) to the recent sedimentary deposits. Moreover, the slump zone is bounded by the Karewa fault on the eastern part of the prospect. The study area is drilled up to a depth of 2,215 m by the well Karewa-1, which has penetrated the targeted Pliocene Manga C1 geologic formation. The Manga C1 is a sand-dominated formation associated with elevated amplitudes on seismic data (Figure 14.2).

The original seismic data is associated with several noises that disturb visualization and hence interpretation of Karewa MTD (Figure 14.3a–b). Structural conditioning of the data removes the noises and distorted reflections, and enhances the targeted zone (Figure 14.3c–d). Internally, the Karewa MTD is observed to be structurally deformed (Figures 14.3 and 14.4). The top of the MTD runs more or less parallel to the upper bounding PPS. The bottom of the MTD i.e. the BSS exhibits concave geometry with the limbs of shearing surface transgressing upwards on the eastern part of the survey area (Figure 14.4).

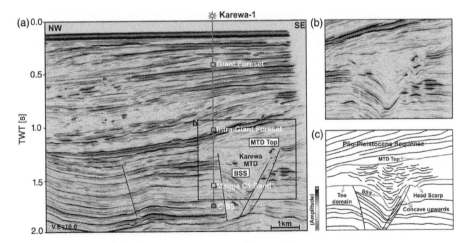

Figure 14.2   (a) Conditioned seismic section for line IL 1278 demonstrating the subsurface architecture of the Karewa prospect, drilled by well Karewa-1, which is terminated at C B5 strata. The eastern part of the prospect is dominated by Karewa MTD. (b) Zoomed view of the MTD zone, as marked by black rectangle in (a). (c) Interpreted sketch of corresponding Karewa MTD, bounded by the Plio-Pleistocene sequence (PPS) on the top and basal shearing surface (BSS) at the bottom, exhibiting concave upward geometry.

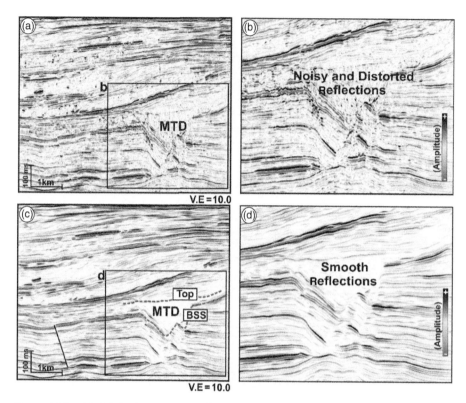

Figure 14.3 (a) Original time-migrated seismic section for line IL 1124 from the Karewa prospect. (b) Zoomed view of the MTD zone, marked by black rectangle in (a). The MTD zone is mixed up with noises and distorted events. (c) DSMF conditioned seismic section for the same line (IL 1124). (d) Zoomed view of the MTD zone, marked by black rectangle in (c) showing improved and smoothed image of subsurface.

The headwall and toe domains of MTD lie on the eastern and western parts of the survey area, respectively. The translation domain, i.e. the main body of the Karewa MTD is bounded by a set of fault systems that are antithetic to the Karewa fault lying on the eastern part (Figure 14.4b, c, e, and f). The MTD is associated with discontinuous reflections, as seen by low similarity attribute (Figure 14.5a) and variable dips, and implied by dip variance attribute (Figure 14.5b). The MTD is not only discontinuous in nature but also internally deformed and contains rafted sediment units, resulting in the loss of energy (Figure 14.5c) and frequency, as evidenced by the variable frequency attribute (Figure 14.5d).

Based on an interpreter's knowledge and skills guided by seismic character- istics and properties associated with MTD, the "MTD-yes" and "MTD-no"

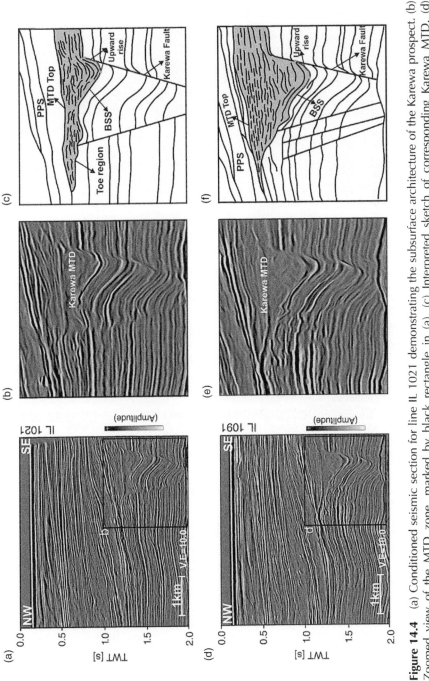

Figure 14.4 (a) Conditioned seismic section for line IL 1021 demonstrating the subsurface architecture of the Karewa prospect. (b) Zoomed view of the MTD zone, marked by black rectangle in (a). (c) Interpreted sketch of corresponding Karewa MTD. (d) Conditioned seismic section for line IL 1091 demonstrating the subsurface architecture of Karewa prospect. (e) Zoomed view of the MTD zone, marked by black rectangle in (d). (f) Interpreted sketch of corresponding Karewa MTD.

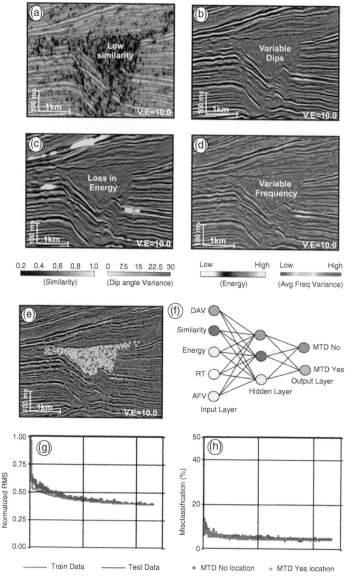

Figure 14.5   Computed seismic attributes (a) similarity, (b) dip angle variance (DAV), (c) energy, and (d) average frequency variance (AFV) within the Karewa MTD, associated with low similarity, variable dips, low energy, and variable frequency content. (e) Chosen locations showing MTD-yes (green dots) and MTD-no (red dots) on a portion of seismic line (IL 1124) as an example. (f) A designed, fully connected MLP network. (g) Normalized RMS error, and (h) misclassification (%) for the train (red) and test (blue) data sets, respectively. RT: Reference Time.

Table 14.1 Relative contribution of each individual seismic attribute to the neural training

Seismic attributes	Weights
Similarity	96.8
Dip angle variance	87.8
Energy	78
Reference time	75
Average frequency variance	62

locations (Figure 14.5e) are picked manually over the randomly selected few-xlines and inlines from the huge 3-D volume. Using these small segments of data, a fully connected multi-layer perceptron (MLP) network (Figure 14.5f) is iteratively trained by a feedforward process, which results in a minimum nRMS error (Figure 14.5g) and low misclassification % (Figure 14.5h) for both the train and test data sets. It is observed that the nRMS of 0.3 and 0.45, and minimum misclassification % of 6.05 and 8.02 % are achieved after 25 iterations for the train and test data sets, respectively. The relative contribution provided by input seismic attributes, while training the system, is given in Table 14.1.

The machine automatically generates the MTDC meta-attribute, which is a probability cube with values ranging between 0 and 1 (bottom panel of Figure 14.6). The values closer to 0 show the least probability of MTD and those closer to 1 indicate the highest probability for the occurrence of MTD. An optimum color scale (i.e., pastel) is used in such a way that the maximum probability is visualized by fixing the threshold value of 0.75 for automatic delineation of MTD by machine as the final outcome, and those pertaining to the minimum probability are made transparent. The comparison between the conditioned seismic section (Figure 14.6a) and the same section co-rendered with the machine-generated MTDC meta-attribute (Figure 14.6b) exhibits that this MTDC meta-attribute has efficiently predicted the Karewa MTD along a line, which was not considered while training the system. This is done as a quality check as a part of neural training. Moreover, the MTDC meta-attribute has also arrested the lateral extension of MTD in the headwall and toe domains lying to the eastern and western parts of the study area. The base of the MTD resembles a w-shape structure with the BSS rising upward (Figure 14.6b). The MTDC meta-attribute has clearly brought out the structural elements and NW-SE elongated 3D geometry of MTD (Figure 14.7). The MTD covers an area of ~20 km^2 and is dominant in the SE of the Karewa prospect. The MTD is internally deformed with a sheared base.

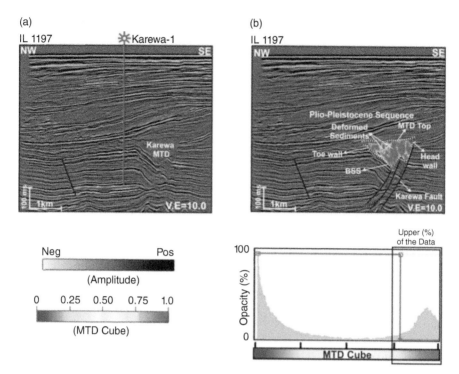

Figure 14.6 (a) Seismic section for line IL 1197 demonstrating the Karewa MTD in the eastern part. (b) The same section, clipped with the machine-generated MTDC meta-attribute. A pastel color scale is used to display meta-attribute, where red signifies the highest probability of MTD.

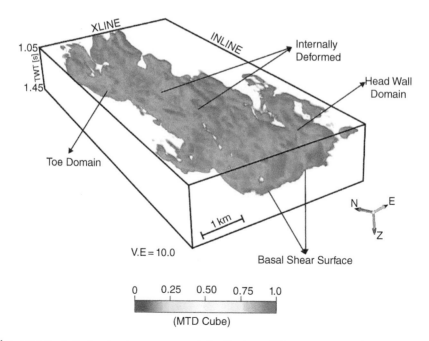

Figure 14.7 3-D structural geometry and distribution of Karewa MTD, as brought out by the MTDC meta-attribute computation from the 3-D seismic data in the study region.

14.4. Summary

The major conclusions drawn from this study are:

- A new AI-based workflow has been designed by which a set of individual seismic attributes corresponding to MTDs has been combined into a single attribute, defined as the MTDC meta-attribute.
- The MTDC meta-attribute has been very efficient in capturing 3-D structural elements of the MTD in the Karewa prospect from seismic reflection data.
- The study brought out 3-D structural configuration of NW–SE elongated MTD covering an area of ~20 km^2. The MTD is dominant in the SE of the Karewa prospect, and internally deformed with a sheared base.
- This approach is fast and semi-automatic and is suitable for advanced interpretation of MTDs in any sedimentary basin of the world, and can be further extended for advanced interpretation of any other geologic feature from 3-D seismic data.

References

Bull, S., Cartwright, J., & Huuse, M. (2009). A review of kinematic indicators from mass-transport complexes using 3D seismic data. *Marine and Petroleum Geology, 26* (7), 1132–1151. https://doi.org/10.1016/j.marpetgeo.2008.09.011

Kumar, P. C., & Sain, K. (2018). Attribute amalgamation-aiding interpretation of faults from seismic data: An example from Waitara 3D prospect in Taranaki basin off New Zealand. *Journal of Applied Geophysics, 159*, 52–68. https://doi.org/10.1016/j.jappgeo.2018.07.023

Kumar, P. C., Omosanya, K. O., & Sain, K. (2019a). Sill Cube: An automated approach for the interpretation of magmatic sill complexes on seismic reflection data. *Marine and Petroleum Geology, 100*, 60–84. https://doi.org/10.1016/j.marpetgeo.2018.10.054

Kumar, P. C., Sain, K., & Mandal, A. (2019b). Delineation of a buried volcanic system in Kora prospect off New Zealand using artificial neural networks and its implications. *Journal of Applied Geophysics, 161*, 56–75. https://doi.org/10.1016/j.jappgeo.2018.12.008

Kumar, P. C., & Sain, K. (2020). A machine learning tool for interpretation of Mass Transport Deposits from seismic data. *Scientific Reports, 10*, 14134, 1–10. https://doi.org/10.1038/s41598-020-71088-6

Martinsen, O. J. (1994). Mass movements. In A. Maltman (Ed.), *The geological deformation of sediments* (pp. 127–165). Chapman and Hall. https://doi.org/10.1007/978-94-011-0731-0

Omosanya, K. O. (2018). Episodic fluid flow as a trigger for Miocene-Pliocene slope instability on the Utgard High, Norwegian Sea. *Basin Research, 30*(5), 942–964. https://doi.org/10.1111/bre.12288

Posamentier, H. W., & Kolla, V. (2003). Seismic geomorphology and stratigraphy of depositional elements in deep-water settings. *Journal of Sedimentary Research, 73*(3), 367–388. https://doi.org/10.1306/111302730367

Tingdahl, K. M. (1999). Improving seismic detectability using intrinsic directionality, Paper B194. *Earth Science Centre, Goteberg University.* https://doi.org/10.1016/S0920-4105(01)00090-0

Varnes, D. J. (1978). *Slope movement types and processes.* Washington, DC: National Academy of Sciences.

APPENDIX A

MATHEMATICAL FORMULATION OF SOME SERIES AND TRANSFORMATION

Fourier Series

The Fourier series is a mathematical expansion of a periodic function, say $f(x)$, according to which the function can be expressed as an infinite sum of sines and cosines.

A function of period M such that $f(x + M) = f(x)$ for all x in the domain of f.

Let us consider a function $f(x)$ such that $-M \leq x \leq M$; the Fourier series is given as

$$f(x) = A_0 + \sum_{n=1}^{\infty} A_n \cdot \cos\left(\frac{n\pi x}{M}\right) + \sum_{n=1}^{\infty} B_n \cdot \sin\left(\frac{n\pi x}{M}\right) \qquad \text{(A-1)}$$

where

$$A_0 = \frac{1}{2M} \int_{-M}^{M} f(x)dx$$

$$A_n = \frac{1}{M} \int_{-M}^{M} f(x) \cos\left(\frac{n\pi x}{M}\right)dx, \quad n > 0$$

$$B_n = \frac{1}{M} \int_{-M}^{M} f(x) \sin\left(\frac{n\pi x}{M}\right)dx, \quad n > 0$$

Meta-Attributes and Artificial Networking: A New Tool for Seismic Interpretation,
Special Publications 76, First Edition. Kalachand Sain and Priyadarshi Chinmoy Kumar.
© 2022 American Geophysical Union. Published 2022 by John Wiley & Sons, Inc.
DOI: 10.1002/9781119481874.appendixA

Fourier and Inverse Fourier Transforms

Fourier Transform (FT) and Inverse Fourier Transform (IFT) are powerful tools for analysing the signals or waveform in different domains. Fourier Transform is applied to convert a function from the time domain to the frequency domain. The reverse of this holds true for the Inverse Fourier Transform.

Let us consider a function $g(t)$. The FT of this function is given as

$$g(t) = G(f) = \int_{-\infty}^{\infty} g(t)e^{-i2\pi ft} \, dt \tag{A-2}$$

where $G(f)$ generates a power spectrum of $g(t)$ that contains the frequency (f). Now let us apply IFT to $G(f)$. Thus, this is given as

$$G(f) = g(t) = \int_{-\infty}^{\infty} G(f)e^{i2\pi ft} \, df \tag{A-3}$$

Hence, equation A3 results back to the original function $g(t)$. Both $g(t)$ and $G(f)$ are called the Fourier pairs.

Hilbert Transform

The Hilbert Transform (HT) is a phase rotation operator that imparts a phase shift of $\pm 90^{\circ}$ to a signal $x(t)$. The Hilbert Transform of $x(t)$ is represented by $\hat{x}(t)$ and is given as

$$\hat{x}(t) = \frac{1}{\pi} \int_{-\infty}^{\infty} \frac{x(k)}{t-k} \, dk \tag{A-4}$$

The inverse Hilbert transform is given as

$$x(t) = -\frac{1}{\pi} \int_{-\infty}^{\infty} \frac{\hat{x}(k)}{t-k} \, dk \tag{A-5}$$

The important properties of a signal $x(t)$ and its HT $\hat{x}(t)$ are that both of them have the same amplitude spectrum and same energy spectral density. Moreover, the signal and its HT are mutually orthogonal to each other.

In seismic data analysis, the application of HT to the seismic trace generates the quadrature component of the trace and finds an important application in complex trace analysis.

Let us consider a cosine wave $x(t)$ with amplitude A, frequency f, and phase θ. Mathematically this can be expressed as

$$x(t) = A \cos(2\pi ft + \theta) \tag{A-6}$$

Thus, applying HT to the signal in equation A6 results in

$$y(t) = h(t) * A \cos(2\pi ft + \theta) = A \sin(2\pi ft + \theta) \tag{A-7}$$

The above equation elaborates the HT for any sort of waveform.

Convolution

Convolution is a mathematical operation which, when carried out, determines the amount of overlap a function, say g, exhibits when it is shifted over another function f. The algebra of Schwartz functions defines convolution as a product of two functions.

It is represented as $[f * g]$. Thus, convolution of two functions f and g within a range $[0, t]$ can be mathematically expressed as:

$$[f*g](t) = \int_0^t f(u)g(t-u)\,du \tag{A-8}$$

Convolution has a very important application in seismic data analysis, where the convolution model of seismic data is used to understand how the seismic trace is generated.

APPENDIX B
DIP-STEERING

Dip-steering is the most modern application for the estimation of dip-azimuth volume, which is also referred as a steering cube (Jaglan et al., 2015; Kumar & Mandal, 2017; Kumar & Sain, 2018; Kumar et al., 2019b, c, d; Tingdahl, 1999; Tingdahl and de Groot, 2003). The deep-steering cube contains seismic dip and azimuth information at every sample position. This is calculated by transforming a sub-cube into the 3D Fourier domain and then estimating the maximum dip with the help of a third-order polynomial curve fitting (Tingdahl, 2003) to the sub-cube around the sample of the highest energy in the Fourier domain. Then a search for the local maxima is made and the corresponding dip-azimuth to the local maxima is set as the output. This operation, when carried over a 3D cube, generates a dip-azimuth volume.

The dip-steering process can be improved by aligning the neighbouring signals (Tingdahl, 1999). Let us consider that similar seismic events are present on two adjacent traces. In such case, the phases of these events are likely to be equal:

$$\theta(t_a) = \theta(t_b) \tag{A-9}$$

where, t_a and t_b are the time of events for trace A and B. Thus, it is reasonable to find out the time t' on both of these traces and to adjust it in such a way that the phase coincides. The same phase can be determined at many locations along the seismic signal. Hence, it is necessary to optimize the limit of t' that can be done by searching for the same phase in the interval, and is given as

$$t'_b \in \left[t_a - \frac{t_w}{2}, t_a + \frac{t_w}{2} \right] \tag{A-10}$$

where, t_w represents the aperture of the search. Even if two t' within the same phase are found twice within the same interval, it is not always true that the one nearest to t is the best choice. Reflectors are not always flat; rather they keep on dipping. Thus for dipping seismic events, the optimal time search can be made by

Meta-Attributes and Artificial Networking: A New Tool for Seismic Interpretation,
Special Publications 76, First Edition. Kalachand Sain and Priyadarshi Chinmoy Kumar.
© 2022 American Geophysical Union. Published 2022 by John Wiley & Sons, Inc.
DOI: 10.1002/9781119481874.appendixB

following the local dip directions, and hence the computed dip helps to determine the search direction of the equal phase. This is expressed as:

$$t_b' \in \left[t_{dip\,b}' - \frac{t_w}{2}, t_{dip\,b}' + \frac{t_w}{2} \right]$$ (A-11)

This process of adjusting the intercept time following the phase of the seismic event is known as phase locking and plays a significant role in the dip-steering process. At the intercept time, the local dip and azimuth are followed to the next trace along the path.

For practical applicability of the dip-steering process, readers are suggested to critically read the published articles (e.g., Tingdahl, 1999; Tingdahl and de Groot, 2003; Tingdahl and de Rooij, 2005; Jaglan et al., 2015; Kumar & Mandal, 2017; Kumar & Sain, 2018; Kumar et al., 2019b, c, d).

APPENDIX C
SOLUTIONS TO TASKS IN CHAPTER 3

Solutions for Seismic Cross-Section Interpretation Tasks (Tasks 1–6)

Task 1

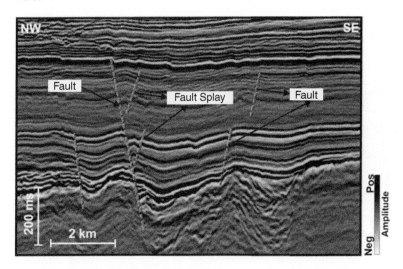

Meta-Attributes and Artificial Networking: A New Tool for Seismic Interpretation,
Special Publications 76, First Edition. Kalachand Sain and Priyadarshi Chinmoy Kumar.
© 2022 American Geophysical Union. Published 2022 by John Wiley & Sons, Inc.
DOI: 10.1002/9781119481874.appendixC

Task 2

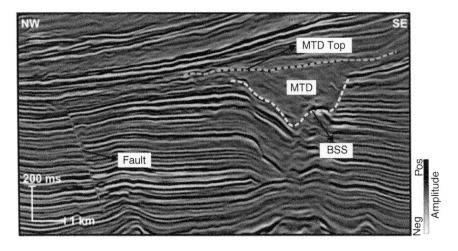

Task 3

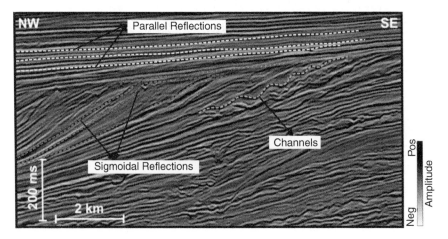

Task 4

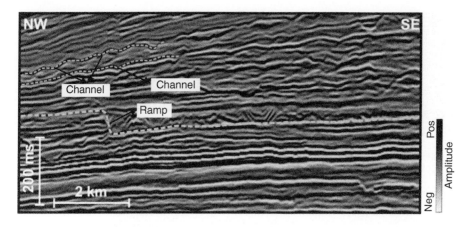

Task 5

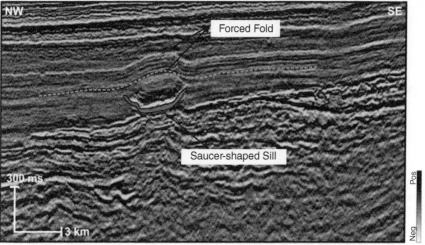

Task 6

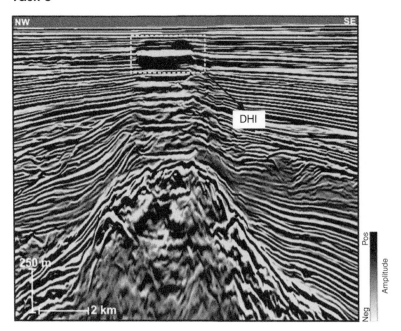

Answers to Numerical Tasks (Tasks 7–10)

Task 7

(a)   Yes, noise bursts are present in the signal.
(b)   Amplitude 5 is a significant noise burst.
(c)   Applying a 3-step median filter to the signal results in x *filtered* (t) = {0, 1, 1, 1, 2, 3}.
(d)   A 3-step median filter shall be an optimized window length.

Task 8

(a)   After convolving the two sequences, the output signal is $y(t)$ = {0, 0, 0, 0, 0, 1, 6}.
(b)   Yes, there is a noise burst in the signal. The noise burst in the output signal is 6.
(c)   After applying a 3-step median filter, the signal becomes y *filtered*(t) = {0, 0, 0, 1}.

Task 9

Refer to Appendix A to understand the mathematical formulation and equations.

Task 10

(a)   The resultant signal after 2D convolution is

$y(t) =$	4	9	11	6
	5	12	16	9
	1	3	5	3

(b)   The basic difference between the convolution and correlation is that convolution rotates the kernel matrix by 180 degrees during the mathematical operation.

(c)   Correlation is used to check the similarity between the sample points in the signal or similarity between the pixels in the image.

INDEX

A

activation function, 79, 80, 82, 85, 120, 136, 156, 173, 188, 218, 238

alpha-trimmed, 98–99

amplitude
  attenuation, 134, 152
  variance, 23, 36, 37, 44

anisotropy, 4

artificial
  intelligence, 63, 74
  neural network, 73, 74, 89, 96, 114, 115, 134, 169, 171, 181, 193, 212, 237

ascent of magma, 212, 223

attenuation, 4, 6, 9, 23, 39, 64, 114, 125, 134, 152

AVO intercept, 4

axon, 75, 77

azimuth, 7, 8, 10, 17, 23, 25, 27, 38, 44, 65, 66, 67, 96, 114, 133, 212, 214, 227, 249, 252

B

backpropagation, 80, 81, 86, 89, 96, 104, 238

bandwidth, 8, 9, 17, 21, 23

basis functions, 85

biological neuron, 76, 78

bipolar sigmoid, 80

buried volcano, 57, 63, 65, 66, 211, 219, 222, 231

C

Caianiello neuron, 78

carbonate reefs, 101

cell, 75, 77

channel complexes, 62, 101

chaotic reflections, 64, 65, 102, 113

chimney cube, 63–64, 113, 114, 128

classification, 76, 86, 87, 102, 106, 238

coherency, 17, 27, 28, 62, 65, 101, 133, 136, 138, 146, 162, 212, 214

complex trace
  analysis, 6, 8, 17, 25, 44
  attributes, 3, 5, 17, 20

connections weights, 75, 78

continuous wavelet transform, 39

co-rendered, 118, 123, 139, 207, 220, 226, 227, 243

cover's theorem, 85

curvature, 7, 8, 10–12, 23, 30–35, 44, 62, 65, 101, 133, 138, 139, 212, 217

D

3D
  cube, 27, 249
  Fourier domain, 27, 249
  reflection seismic data, 169, 185
  seismic data, 6, 61, 63, 67, 73, 74, 89, 108, 113, 129, 148, 169, 181, 193, 211–212
  seismic volume, 27, 95, 180, 231, 235
  time-migrated, 134, 173, 223
  visualization, 27, 113, 207

Meta-Attributes and Artificial Networking: A New Tool for Seismic Interpretation, Special Publications 76, First Edition. Kalachand Sain and Priyadarshi Chinmoy Kumar. © 2022 American Geophysical Union. Published 2022 by John Wiley & Sons, Inc. DOI: 10.1002/9781119481874.index